WATER POLLUTION

Disposal and Reuse

(in two volumes)

VOLUME 1

WATER POLLUTION

DISPOSAL AND REUSE

(in two volumes)

Volume 1

by J. E. ZAJIC

University of Western Ontario
London, Ontario, Canada

MARCEL DEKKER, INC. New York 1971

MARCEL DEKKER, INC.

95 Madison Avenue, New York, New York 10016

LIBRARY OF CONGRESS CATALOG CARD NUMBER 70-163919

ISBN 0-8247-1815-1

PRINTED IN THE UNITED STATES OF AMERICA

This book is dedicated

to

Elizabeth Shane Zajic

PREFACE

The interdisciplinary sciences are unquestionably the most difficult fields to teach and for the student to learn. Environmental science, which includes the study of water pollution, requires the development of talents and skills in many areas. In examining air, water, and soil pollution, or in evaluating natural resources and the population problem, no single subject can be said to be paramount. The student, researcher, or engineer must draw on a knowledge of chemistry, biology, mathematics, statistics, and hardware to solve problems involving the lithosphere, biosphere, atmosphere, and hydrosphere.

If all resources are reduced to the life supporting level, water, air, and energy are most essential. Water of varying degrees of purity is required by every form of life. Water makes up 85 per cent or more of the weight of most plants and animals. It follows that, where fresh water is a limiting factor, life forms are limited. No single compound or solvent has the diverse properties associated with this solvent. Since life processes are so dependent upon it, one cannot visualize life in space without an adequate supply of moisture.

These volumes do not treat the hydrosphere except as that subject relates to purification. Most water purification processes have been going on in nature from that distant point in time at which water became a dominant factor in the surface structure of the planet Earth. Distillation, condensation, filtration, freezing, thawing, and biological degradation of organic contaminants, and even flocculation and clarification, are actually natural phenomena.

In most instances, man has had to utilize these processes as arts, but gradually, through study, they are being reduced to a science. Water pollution control is, however, still an empirical field. It is in great need of basic study and much of the theory still remains to be established.

These volumes are written for anyone interested in water and its purification. Since any chemical in concentrated form in water can be toxic, the work starts with three chapters which lay a predicate for the prime toxic chemicals which man has encountered. Water quality is discussed from both a chemical and biological base. Since the study of water pollution has already drawn together certain terminology, this too must be introduced to the environmentalist at an early stage.

The remaining chapters are placed under either the biological (Vol. 1) or the chemical (Vol. 2) section. Where any biological operation is involved in contamination or purification, it is handled in the biological section. One finds, of course, that not all chapters fit neatly within the classifications used, and where conflict is encountered, my choice of handling is evident.

Likewise, not all the subjects handled in Vol. 2 are chemical. The Information Reliability chapter could have been placed in the first section, however, I believe it should be developed after one has acquired some experience with pollution problems. Similarly, settling reactions can be purely physical, and the same can be said for freezing and thawing; but these topics are discussed in the chemical section. An adequate discussion of the equipment used in water pollution would alone comprise a book, and only a few highlights are covered here. If one desires to be critical, there are many topics which are excellent game. I hope that this work will, nevertheless, contribute to the advancement of water pollution control from an art to a fully developed science.

London, Ontario, Canada

J. E. Zajic
Ph.D., M.S., P. Eng.

ACKNOWLEDGMENTS

As with any book usually an author has a large list of people who have assisted or contributed in some manner in the preparation. My graduate students in water pollution engineering and I have spent hours discussing all the topics herein and many have spent hours looking up material to add to our fund of knowledge. Those that have contributed most are P. R. Toews, T. Constantine, M. Schwartz, Miss Y. K. Ho, K. S. Ng, A. V. Giffen, G. Webster, O. M. Behringer, L. Bithel, N. Lefebvre, D. G. Maclean, R. H. Peters, T. E. Rattray, G. W. Schindel, I. Szigethy, G. M. Wong Chong, A. V. Bell, L. E. C. Jacobs, J. E. Kilotat, E. Kotyk, M. A. Rychlo, and P. G. Shewchuk.

CONTENTS

WATER POLLUTION

Disposal and Reuse
(in two volumes)

VOLUME 1

Chapter 1

WATER QUALITY

I. INTRODUCTION

Historically, the availability of good, potable water has been one of the major factors influencing the development of civilization. In probably the first recorded use of treatment facilities the ancient Chinese and Egyptians used crude methods of chemical coagulation to purify their water. All early methods were patterned after natural water purification systems occurring in nature.

Modern treatment had its beginning with the slow sand filter in England in the early 19th century. It was not until the latter part of the same century, however, that diseases of epidemic proportion were traced to water supplies. Since that time the activities of many thousands of scientists and engineers have been directed toward the development of water purification

1

processes aimed at the elimination of adverse physical, chemical, and pathogenic constituents.

The best compendium on water quality criteria to date has been prepared by McKee and Wolf (1963). It not only gives the criteria for all of the States in the United States but also covers the Pollution Control Council of the Pacific Northwest, New England Interstate Water Pollution Control Commission, the International Joint Commission, etc. Water quality objectives for Ontario are discussed by Van Fleet (1968). Van Fleet's presentation is an excellent guide.

Other important documents relating to water management and legislation are:

1. Canada Water Act (1970),
2. Alternatives in Water Management (1966),
3. Recommended State Legislation and Regulation (1965),
4. Water Resources in Canada, reports 1-5 (1968).

In the United States the Water Quality Act of 1967 (PL 89-234), an amendment of the Federal Water Pollution Control Act (PL 84-660), requires each state to establish water quality criteria for all interstate waters and to develop a plan for the implementation and enforcement of the criteria. In general, the standards must be such as to enhance the quality of natural waters for their "... use and value for public water supplies, propagation of fish and wildlife, recreational purposes, agricultural, industrial, and other legitimate uses. Numerical values are stated for quality characteristics where available and applicable. Biological or bioassay parameters are used, where appropriate" (Federal Water Pollution Control Federation, 1966).

To introduce the environmental engineer to some of the desired objectives in controlling water pollutants, some of the water quality criteria used in Canada and the United States are presented here. "Water Quality Objectives," guarantee a healthy populace and high rate of industrialization. Ontario's objectives were first established in 1964 and updated by amendment in 1967. Water quality criteria can generally be classified under the heading of bacteriological, physical and chemical.

II. BACTERIOLOGICAL CHARACTERISTICS

Coliform bacteria are regarded as indicators of fecal pollutions and of the presence of human pathogens. There are two methods in general use in the laboratory examination of samples for bacteriological quality; one is the most probable number (MPN) technique, used by the Department of Health, and the other the membrane filter (MF) technique. Details of these are treated separately.

A. Interpretation of Results

There is, however, a slight difference in the interpretation of the significance of actual numbers of coliforms indicated in each 100 milliliter (ml) portion. None of the samples having coliform organisms should have an MPN index greater than 10 per 100 ml, while in the MF technique none of the coliform counts should be greater than 4 per 100 ml. If samples of water approach or exceed these limits in consecutive examinations, then an immediate investigation should be initiated to locate and eliminate the source of the contamination. At the same time a series of "special samples" are required to determine the extent of contamination and the progress being made toward its elimination from the water supply. This special sampling continues until the bacteriological water quality proves to be satisfactory. "Special samples" are also required when more than 10% of the samples collected per month and tested by either the MPN or MF methods show the presence of coliform organisms.

A third method of analysis involves running a series of confirmation tests to confirm and definitely establish the presence or absence of a variety of pollution indicator bacteria. These specialized tests are required to identify pathogenic cultures as to genus and species.

B. Frequency of Sampling

Contamination is often intermittent and may not be revealed by the examination of a single sample. The examination of a single sample can indicate no more than the conditions prevailing at the time of sampling; a satisfactory result cannot guarantee that the observed conditions will prevail in the future. A series of samples over a period of time is thus required.

To ensure reliable results, samples should arrive at the testing laboratory within 24 hr of sampling or be refrigerated if delay is unavoidable. The sample should be collected directly into sterile bottles and not via a dipper or some other container. It should be stressed that the reliability of the results is wholly dependent upon the employment of proper sampling techniques and the care with which the samples are collected. The minimum number of samples required from specific sources are shown in Table 1.

TABLE 1

SAMPLING FREQUENCY

Description of source	No. of samples	Minimum frequency of sampling
Treated surface water	1 raw and 1 treated at plant	Once per week
Treated ground water	1 raw and 1 treated from each source	Twice per week
Untreated ground water	1 raw from each source and 1 from each point of entry into the distribution system	Once per week

The minimum number of samples to be collected and the frequency of sample collection from a distribution system are often based on the size of the population served (Table 2).

TABLE 2

SAMPLE FREQUENCY BASED ON POPULATION

Population served	Minimum number of samples per month	Minimum frequency of sampling intervals
Up to 1,000	2	Twice per week
1,001–100,000	10 + 1 per 1000 of population per month	Once per week
Over 100,000	100 + 1 per 10,000 of population per month	Once per day

The number of samples determined with the use of the above table should not include plant effluents whether treated or otherwise. The frequency and size of sampling should follow the guide adopted by the U.S. Public Health (1962) shown in Fig. 1.

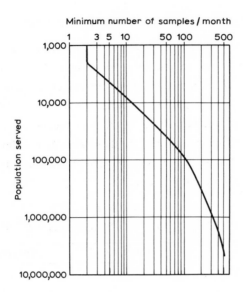

FIG. 1. Flow chart of activated sludge process.

As a typical example let us consider a municipality of 2400 persons with two wells providing treated ground water to the system. Bacteriological sampling would consist of the submission twice per month of a raw water sample from each of the sources of supply and a treated water sample from each of the points of discharge to the distribution system. In addition, a total of 12 bacteriological samples per month would have to be taken from various points in the distribution system. This would entail the submission of three samples per week in order to meet the sampling frequency requirement.

The responsibility for taking the required number of samples lies with the operating authority, whether it be a municipality or an individual who owns a private water supply. The total number of samples to be collected monthly may be examined by the Water Resources Commission, other

government laboratories, water works authorities, or by commercial laboratories. "Special samples" are not included in the total number of samples
required above.

III. PHYSICAL CHARACTERISTICS

Physical tests do not directly measure the safety of a water supply;
however, they do give an indication of its acceptability. Thus, objectives
governing the physical characteristics of the water are somewhat less
stringent than those required for bacteriological control. The physical
qualities of concern are turbidity, color, taste and odor, temperature and
pH. [Except for temperature and pH, these results may be reported in parts
per million, ppm (Table 3).] However, oftentimes other units have become
established and accepted in practice.

TABLE 3

PHYSICAL QUALITY OF TREATED WATER [a]

Parameter	Objective	Acceptable Limit
Color, TCU [b]	< 5	15
Odor, TON [c]	0	4
Taste	Inoffensive	Inoffensive
Turbidity, JTU [d]	< 1	5
Temperature, $^{\circ}$C	<10	15
pH [e]	--	6.5-8.3

[a] To be examined according to the latest edition of Standard Methods
for the Examination of Water and Wastewater (American Public Health
Association, American Water Works Association, and Water Pollution
Control Federation), or other acceptable methods as approved by the
control agency.

[b] True Color Unit, platinum-cobalt scale.

[c] Threshold Odor Number.

[d] Jackson Turbidity Unit.

[e] Has significance in controlling corrosion and scaling tendency of the
water.

A. Turbidity

Turbidity should average not more than 1 (turbidity) unit, although a Jackson turbidity unit of 5 is acceptable. At levels approaching 10 units the water may appear cloudy to the observer. Plants that provide complete treatment should routinely produce water than meets this objective. Ground water supplies will normally meet the objective without the need for treatment.

B. Color

Color should average not more than 5 (apparent color) units. Color does not occur too frequently in the natural waters, particularly in Canada. Leaching effect of the water on organic material found in certain watersheds may range beyond 50 units. Removal is possible with alum coagulation, sedimentation, and filtration.

C. Temperature

Temperature is a physical characteristic about which little can be done. The most desirable range is from 40-50°F (4-10°C). Higher temperatures tend to make water less palatable and reduce its suitability for air conditioning purposes. Temperatures above 80°F (27°C) are unsuitable and above 90°F (32°C) are unfit for public use.

IV. CHEMICAL CHARACTERISTICS

Under normal circumstances, analyses for chemical constituents need only be made semiannually. If, however, the supply is suspected of containing undesirable elements, compounds, or materials, then periodic determinations for the suspected toxicant or material should be carried out at more frequent intervals. Where experience indicates that particular substances are consistently absent from a water supply, semiannual examinations for these substances may be omitted on government approval.

A. Limits for Chemical Constituents

The chemical constituent concentrations in water may be broken down into two categories: those concentrations that can be tolerated if another, more suitable source is not available, and those concentrations that

constitute grounds for rejection. Because of the rapidly changing field of
complex chemicals, it has become important that the objectives for chemical constituents be reviewed regularly.

The chemical substances shown in Table 4 should not be present in a
water supply in excess of the listed concentrations where, in the judgment
of a Water Resources Commission, other more suitable supplies are or can
be made available.

TABLE 4

GENERAL TOLERANCE LIMITS OF CHEMICAL WATER POLLUTANTS [a]

Substance	Concentration, mg/l or ppm	
	Acceptable	Objective
Alkyl benzene sulfonate (ABS)	0.5	
Ammonia (as N)	0.5	< 0.01
Arsenic (As)	0.01	$< 0.75 \times 10^{-2}$
Calcium (Ca)	200	
Chloride (Cl)	250.0	<250
Copper (Cu)	1.0	< 0.01
Carbon chloroform extract (CCE)	0.2	< 0.05
Cyanide (CN)	0.01	
Fluoride (F)	_[b]	
Iron (Fe)	0.3	< 0.05
Magnesium (Mg)	150.0	< 50
Manganese (Mn)	0.05	< 0.01
Mercury (Hg)	0.05	< 0.05
Methylene blue active substances	0.5	< 0.2
Nitrate (NO_3)	45.0	
Phenols	0.002	N.D. [c]
Phosphate (PO_4)	0.2	< 0.2
Sulfate (SO_4)	250.00	
Sulfide (H_2S)	.05	<.05
Total dissolved solids	1000.0	<500

TABLE 4

GENERAL TOLERANCE LIMITS OF CHEMICAL
WATER POLLUTANTS [a] (Continued)

Substance	Concentration, mg/1 or ppm	
	Acceptable	Objective
Total hardness as $CaCO_3$	180	120
Uranyl ion as UO_2	5.0	1.0
Zinc (Zn)	5.0	1.0

[a] The presence of substances in excess of the concentrations listed in this
table should constitute grounds for rejection of the supply.

[b] Discussed separately.

[c] N.D. (not detectable).

B.   Taste and Odor

Taste and odor are very closely related and are caused by the same
conditions. A good objective for odor is a threshold odor number (TON)
not greater than 3 and the taste should not be objectionable. Common
sources of taste and odor are:

1.   Dead or decaying organic matter.
2.   Living organisms and oils from algae.
3.   Industrial wastes (phenolic wastes may produce a medicinal taste and
     odor at concentrations as low as 1 ppb when contacted by chlorine).
4.   Dissolved gases (hydrogen sulfide, methane, etc.).
5.   Dissolved minerals (chlorides, sulfates, and metallic salts, such as
     copper and iron).

The physical characteristics provide only part of the picture of water
quality. They are, however, very important when related to surface waters
of variable quality. Normally, ground water will meet all physical requirements, but occasionally the taste and odor of a well or spring supply will be
significant. The presence of hydrogen sulfide gas in a well supply in any
amount will usually discourage its use. Algae blooms produce off-flavors,
which are becoming a greater nuisance as lake waters are being used more
frequently in potable water production.

Consumer acceptance is an excellent guide as to whether taste and odor criteria are being met.

Chemical standards for acceptable water have been scrutinized for over 60 years. In other communities and water plants, semiannual examination often suffices to establish whether objectives are being met. Frequency of examination is often set by regulatory agencies.

Assays for common toxic chemicals must be completed. Common toxicants are arsenic, barium, boron, cadmium, chromium, cyanide, lead, nitrate (nitrite), selenium and silver. Objective, acceptable, and maximum permissible limits are shown in Table 5.

Many synthetic and highly toxic biocides are produced and find their way into raw water. A 24-hr composite sample of raw water should be collected and examined at least once every three months for the common biocides. These are aldrin, chlordane, DDT, dieldrin, endrin, heptachlor, heptachlor epoxide, lindane, methoxychlor, organic phosphates + carbamates, toxaphene, and herbicides such as 2, 4-D, 2, 4, 5-T and 2, 4, 5-TP. Limits for these are summarized in Table 6. Choline esterase inhibition is used to determine toxic biocide levels.

C. Problems Associated with Chemical Constituents

1. Alkyl Benezene Sulfonate (ABS)

Contamination of drinking water supplies with ABS results from its disposal as a detergent in household and industrial wastes. The concentration of ABS in municipal sewage ranges as high as 10 ppm. Linear alkylsulfates (LAS) which have replaced many ABS compounds are found at the same level. Such contamination may appear in both surface and ground water supplies. The objective for either ABS or LAS in water supplies has been set at 0.5 ppm (mg/liter). Tests have confirmed that 1 ppm presents an off-taste to water. This level often causes foaming. No apparent toxic effects are evident with water containing 50 ppm of ABS. Linear alkylsulfonates should be handled in a manner similar to ABS compounds until specific control levels are established.

TABLE 5

DRINKING WATER STANDARDS FOR TOXIC CHEMICALS

Toxicant	Objective mg/liter	Acceptable limit, mg/liter	Maximum permissible limit, mg/liter
Arsenic (As)	Not detectable[a]	0.01	0.05
Barium (Ba)	Not detectable	< 1.0	1.0
Boron (B)	--	< 5.0	5.0
Cadmium (Cd)	Not detectable	< 0.01	0.01
Chromium (Cr^6)	Not detectable	< 0.05	0.05
Cyanide (CN)	Not detectable	0.01	0.20
Lead (Pb)	Not detectable	< 0.05	0.05
Mercury (Pb)	Not detectable		0.05
Nitrate – Nitrite (N)	<10.0	<10.0	10.0
Selenium (Se)	Not detectable	< 0.01	0.01
Silver	--	--	0.05

[a]Not detectable by the method described in the latest edition of "Standard Methods" (APHA, AWWA, & WPCF), or by any other acceptable method approved by the control agency.

TABLE 6

RAW AND DRINKING WATER STANDARDS FOR BIOCIDES[a]

Biocide[b]	Objective and acceptable limits	Maximum permissible limits, mg/liter
Aldrin	Not detectable[c]	0.017
Chlordane	Not detectable	0.003
DDT	Not detectable	0.042
Dieldrin	Not detectable	0.017
Endrin	Not detectable	0.001
Heptachlor	Not detectable	0.018
Heptachlor epoxide	Not detectable	0.018
Lindane	Not detectable	0.056
Methoxychlor	Not detectable	0.035
Organic phosphates + carbamates[d]	Not detectable	0.100
Toxaphene	Not detectable	0.005
Herbicides (e.g., 2,4-D, 2,4,5-T, 2-4,5-TP)	Not detectable	0.100

[a]NOTE: Maximum permissible limits are adopted from the Report of the National Technical Advisory Committee on Water Quality Criteria to the Secretary of the Interior, U.S. Department of the Interior, 1968. These limits are to be regarded as tentative standards since they are still being evaluated.

[b]Conventional water treatment has little effect on these dissolved biocides.

[c]Not detectable by an acceptable method of analysis as approved by the control agency.

[d]Expressed as parathion equivalents in cholinesterase inhibition.

2.   Arsenic (As)

The widespread use of inorganic arsenic in insecticides has resulted in its presence in animal foods, tobacco, and other sources. This makes it necessary to set a limit on the concentration of this chemical in drinking water. Our present knowledge concerning the potential health hazards associated with the ingestion of organic arsenic indicates that the concentration of arsenic in drinking water should not exceed 0.01 ppm and concentrations in excess of 0.05 ppm (mg/liter) are grounds for rejection.

3.   Barium (Ba)

Barium is not a common contaminant of water. An excess of 1.0 ppm is grounds for rejection because of the seriousness of the toxic effects of barium on the heart function, blood vessels, and nerves.

4.   Boron (B)

Boron is rapidly absorbed by the human intestine and is excreted in the urine. Ingestion of large amounts affects the central nervous system. Protracted ingestion results in a clinical syndrome referred to as "boroism." The maximum permissible limit in water is 5.0 mg/liter.

5.   Cadmium (Cd)

Cadmium has high toxic potential. Very little attention has been paid to this constituent.

Seepage of cadmium into ground water from electroplating plants has resulted in concentrations up to 3.2 ppm. Other sources of cadmium contamination in water arise from zinc-galvanized iron in which cadmium is used. Tests have shown that concentrations up to 0.01 ppm can be tolerated, but concentrations in excess of this are considered grounds for rejecting a water supply.

6.   Carbon Chloroform Extract (CCE)

The carbon chloroform extract (CCE) test is a practical measure of water quality and is a safeguard against the intrusion of excessive amounts of potentially toxic material into water. It is proposed as a technically practical procedure that provides a measure of protection against the presence of undetected toxic materials.

The test is an indication of organic material in the treated water, the presence of which shows that pollutants have not been removed in the treatment process. The objective for CCE is 0.2 ppm.

7.   Chloride (Cl)

The objective for chloride has been set at an upper limit of 250 ppm. Above this level a salty taste is apparent. Many municipalities experience higher levels of chloride with varying taste intensities. Abnormal amounts of chloride in a natural water suggest pollution probably of a chemical origin.

8.   Chromium (Cr^{6+})

Chromium is another unnatural constituent of water supplies and is not known to be either an essential or beneficial element in the body. Its presence is indicative of industrial pollution probably caused by a plating or tannery operation. Concentration in excess of 0.05 ppm as hexavalent chromium (Cr^{6+}) is grounds for rejection. Trivalent chromium is not believed to be toxic.

9.   Copper (Cu)

Copper is an essential and beneficial element in human metabolism. A deficiency in copper results in nutritional problems in infants. Copper imparts some taste to water and is detectable in ranges from 1 to 5 ppm. Small amounts are not generally regarded as toxic, but very large doses may cause sickness and in extreme cases liver damage. In the use of copper sulfate in a surface water supply for algal control, the levels must be closely controlled.

Since copper in small amounts does not constitute a health hazard but imparts an undesirable taste to drinking water, a reasonable level of 1.0 ppm is a recommended objective.

10.   Cyanide (CN)

Since proper treatment of water will reduce cyanide levels to less than 0.01 ppm, this objective is usually adopted. For the protection of health, concentrations above 0.2 ppm constitute grounds for rejection. A substantial safety factor is provided at the 0.01 ppm concentration because of the rapidly fatal effect of this chemical.

11. Fluoride (F)

Fluoride in drinking water prevents dental caries in children and to a lesser degree in young adults. Where the addition of fluoride to the water supply is practiced, a 1.0 ppm concentration is recommended, with a permissible operating range of 0.8 - 1.2 ppm. It is believed that mottling of the teeth or enamel fluorosis occurs at concentrations above 1.2 ppm. When fluoride is naturally present, the concentration should not average more than 1.2 ppm. Presence of fluoride in concentrations more than 1.5 ppm should be rejected. In the arctic and subarctic areas fluoride should be maintained at 1.4 ppm.

12. Iron (Fe)

Iron is at times a highly objectionable constituent in water for either domestic or industrial supplies. The domestic consumer complains of the brownish color that iron imparts to plumbing fixtures and laundered goods. Iron also affects the taste of beverages. The upper limit for iron is 0.3 ppm, with the overall objective being set at 0.05 mg/liter. Water mains may become fouled by the masses of stringy growths associated with iron oxidizing microbes. Associated with iron is the corrosion and encrustation problem. The American Water Works Association has established a 90-day encrustation rate on stainless-steel coupons which should not exceed 0.05 mg/cm^2. Galvanized iron should not exceed 5.0 mg/cm^2.

13. Lead (Pb)

Lead taken into the body can be seriously injurious to health or even lethal. Lead should not exceed 0.05 ppm. Concentrations in excess of this amount are grounds for rejection.

14. Methylene Blue Active Substances (MBAS)

This determination indicates the presence of synthetic detergents, foaming, excessive turbidity, odor containing substances, etc. The effectiveness of chlorine disinfection decreases if the MBAS is over 4.0 mg/liter of ABS-equivalent. A concentration of MBAS above 0.5 mg/liter as ABS-equivalent is indicative of other wastewater pollution.

15. Uranyl Ion

Uranyl ions produces damage to kidneys. The threshold limit for taste is 10 mg/liter as UO_2, which is less than the safe limit for ingestion. The maximum permissible limit is 5.0 mg/liter, based on color and taste considerations.

16. Manganese (Mn)

Manganese presents much the same nuisance conditions as iron. It is difficult to remove this chemical and it is recommended that concentrations not exceed 0.05 ppm.

17. Nitrate (NO_3) and Nitrites (NO_2)

Serious and occasionally fatal poisonings of infants have occurred following ingestion of well waters shown to contain high levels of nitrate. Wastes from chemical fertilizer plants and field fertilization run-off are sources of such pollution.

Nitrate poisoning appears to be confined to infants during their first few months of life. Nitrates cause infantile methemoglobinemia ("blue-baby" condition). It can be cured or terminated by providing nitrate-free water. Nitrates cause irritation of the mucuous membranes of the stomach in adults and increased diuresis.

Several authorities have adopted the objective of <10.0 ppm nitrate as NO_3 with an upper acceptable limit of 45.0 ppm. In areas where the nitrate content of water is known to be in excess of the listed concentration, the public should be warned of the potential dangers of using the water for infant feeding.

18. Phenol

Undesirable tastes often result from the chlorination of waters containing extremely low concentrations of phenol. The objective for phenol is 2 ppb (0.002 ppm).

19. Selenium (Se)

Levels of selenium in excess of 0.01 ppm constitute grounds for rejection of a water supply. Trace amounts of this chemical are essential to man, while higher concentrations appear to be extremely toxic in a manner similar

to arsenic. Surveys have also shown that selenium increases the rate of
dental caries in permanent teeth.

20. Silver (Ag)

A water supply should be rejected if it contains more than 0.05 ppm of
silver. This level is established, not because of toxic effects, but due to
the unsightly, permanent blue-grey discoloration of the skin, eyes, and
mucuous membranes which results from its ingestion. Evidence indicates
that silver, once absorbed, is bound indefinitely in the tissues, particularly
the skin.

21. Sulfate (SO_4)

A diarrhea effect is commonly noted by newcomers and casual users of
waters high in sulfates. Persons adapt to these waters in relatively short
time periods. The taste of the water is adversely affected at the upper limit,
recommended as 250 ppm.

22. Total Dissolved Solids

High dissolved-solids concentrations are associated with correspondingly high levels of sulfates and/or chlorides. An upper limit of 500 ppm has
been set in order to control undesirable taste and diarrhea.

23. Zinc (Zn)

Zinc is an essential and beneficial element in human metabolism and
does not appear to have a serious effect on health. Zinc salts have a tendency to impart a milky appearance to water at 30 ppm and a metallic taste
about 40 ppm. These are the only apparent undesirable characteristics.
An objective of 5.0 ppm is acceptable.

D. Miscellaneous Chemical Considerations

In addition to the purely chemical characteristics that should be considered, there are other characteristics, which for want of a better term,
may be classified as miscellaneous.

1. Acidity and Alkalinity

Alkalinity and acidity of water refer to the amounts of acids or bases
present and are measured in ppm. These are not to be confused with pH,

which is a measure of the hydrogen ion concentration measured on an arbitrary scale from 0 to 14. There are no particular limits for either alkalinity or acidity and both are expressed in terms of $CaCO_3$. Highly acid or alkaline waters should be avoided and could be dangerous.

Acidity is prevalent in many northern waters, while alkalinity is a characterisitic of waters found in southern Ontario. Acidity is not desirable in a municipal water system primarily because it increases corrosion.

Alkalinity refers to the carbonate, bicarbonate, and hydroxide content of a water and is commonly found in the form of a carbonate of soda (Na) and as bicarbonates of calcium (Ca) and magnesium (Mg). Where the alkalinity exceeds the hardness, the presence of basic salts, generally sodium (Na) and potassium (K), is indicated. If the alkalinity is less than the hardness, then salts of Ca^{2+} and Mg^{2+} are present in association with sulfates, chlorides, or nitrates. Grains of hardness for drinking water should be maintained between 4.7 and 7.0, or less than 81-120 mg/liter of $CaCO_3$. Very good waters have less than 80 mg/liter of $CaCO_3$.

2. Carbon Dioxide

In surface supplies the normal CO_2 content will range from 0.5 to 2.0 ppm, while in ground water it will range as high as 50 ppm. A proper balance of carbon dioxide in water will ensure that the water is neither corrosive nor scale-forming.

3. Hardness

Water ranges from less than 10 ppm to 1800 ppm. A preferable hardness is in the range 90-100 ppm. Above 500 ppm the water may be considered objectionable for domestic use. Waters with hardness less than 30 ppm are quite soft and usually not corrosive.

4. Hydrogen Sulfide

Even trace amounts of hydrogen sulfide (H_2S) will create a taste and odor characteristic. It is not harmful from a health standpoint at levels that persons would consider for drinking.

5. pH

Natural waters generally range from pH 5.5 to 8.6. Waters with lower pH tend to cause corrosion, and in many cases an upward adjustment to the neutral range (pH 7.0) is necessary. Drinking water with a pH range from 6.5 to 8.3 is necessary. At higher pH values the effectiveness of chlorination decreases, i.e. the rate of kill of microbes decreases.

6. Phosphate

In a natural, unpolluted water, phosphates have little significance. However, due to the increased use of detergents and commercial fertilizers, phosphates are being discharged into lakes and streams in concentrations which greatly affect biological activities in these bodies of water. Consequently, they exert secondary effects on water supplies, which may necessitate the provision of additional treatment facilities. On the other hand, complex phosphates are often introduced into sources of supply for the prevention of corrosion and scaling in water distribution systems.

7. Radiological Limits

The exposure of humans to radiation is viewed as harmful. Exposure to ionizing radiation should be controlled and monitored. Concentrations exceeding the average values presented in Table 7 for a period of one year should constitute grounds for rejection. All samples for analysis from water supply systems should be composited over a three-month period and discharges receiving radioactive wastes should be examined several times per year.

TABLE 7

RADIATION LIMITS

Radionuclides	Concentration, $\mu\mu$Ci/liter
Radium-226 (^{226}Ra)	3
Strontium-90 (^{90}Sr)	10
Gross β-activity (^{90}Sr and α-emitters absent[a])	1000

[a]Absent is taken here to mean a negligibly small fraction of the above specific limits, where the limit for unidentified α-emitters is taken as the listed limit for ^{226}Ra.

TABLE 8
SUGGESTED RAW WATER QUALITY CRITERIA

Ion, organism or substance	Water use			
	Municipal	Industrial	Agricultural	Recreation fish wildlife
Algae, std. units vol.	1000 std. units vol.	1000	absence of toxic algae	
Alkalinity, ppm	120	50–150	--	--
Ammonia, ppm	0.1	--	--	0.3
Biochemical oxygen demand, ppm	max. 1.0–3.0 av. 0.75–1.5	--	--	see DO
Bicarbonates, ppm	150	3–100	--	--
Boron, ppm	1	--	0.2–0.5	--
Cadmium, ppm	0	0	0	0
Calcium, ppm	--	--	40	--
Carbon dioxide, ppm	--	--	20–40	--
Carbonates, ppm	--	200–400	10	--
Chloride, ppm	50	20–250	100	--
Chromium, ppm	0	0.05	0	10.0
Coliform–MPN/100 ml	50	100–1000	--	500–100
Color	20–70	100–1000	--	--
Copper, ppm	3.0	--	0.2	1.0
Cyanides, ppm	0	0	0	0
DO, ppm	over 4.0	0.2–2.0	0.2	over 4

Floating solids	0	0	0	0
Fluoride, ppm	1.5	1.0	--	5.0
Hydrogen ion concentration, pH	6.5-8.5	6.0-9.6	--	5.0-9.5
Iron, ppm	0.3	0.5	--	--
Magnesium, ppm	--	--	20[a]	--
Nitrate, ppm	10	--	--	44
Oil, ppm	0	0	0	0.3
Phenolic compounds, ppm	0	.001-.010	.005-.020	0.2-1.0
Radioactivity μ Ci/liter	background	background	background	background
Sodium, ppm	0-10	50	10	--
Sulfates, ppm	250	100-250	190	--
Suspended solids, ppm	--	--	absence of sludge deposits	
Total dissolved solids, ppm	1000	100-1000		
Temperature, °F (desirable)	50	60	60	32-95
Total nitrogen, ppm	10	--	--	10
Turbidity, ppm	.0-40	10-50	--	2000

[a] SAR values show the interrelationship.

Where the total intake of ^{226}Ra and ^{90}Sr from all sources has been determined, the limits may be adjusted so that the total intake of ^{226}Ra and ^{90}Sr will not exceed 7.3 micro microcuries ($\mu\mu$Ci) per day and 73 $\mu\mu$Ci/day, respectively. When mixtures of ^{226}Ra and ^{90}Sr, and other radionuclides, are present, the above limiting values shall be modified to ensure that the combined intake is not likely to result in radiation exposure in excess of the Radiation Protection Guides recommended by the United States Federal Radiation Council. Radiation limits in certain mining operations, e.g., uranium, are somewhat higher. The radioactivity limits for drinking water in Canada are: acceptable -- 1/3 of the International Commission in Radiological Protection (FCRP) of the maximum permissible concentration in water (MPC)$_w$ for a 168-hr week; maximum permissible limit -- the ICRP(MPC)$_w$ for 168-hr week; objective -- 1/10 of the ICRP(MPC)$_w$ for 168-hr week.

The recommended MPC in effect for some years has been 100 pCi/liter (picocuries or micromicrocuries/liter of air) of radon gas in equilibrium with short-lived disintegration products; the latter are RaA (polonium-218), RaB (lead-214), etc. This MPC has been documented officially in the Brit. J. Radiol. and also appears in several American publications.

The lower MPC subsequently recommended by the ICRP has not received general acceptance, although it has been partly confirmed by the latest MCP sepcified by the American Standards Association of 1.3 x 10^5 MeV of α-energy "released by the decay through RaC" of 100$\mu\mu$Ci of each of the three daughters (RaA, RaB, and RaC)." The IAEA standards, provisionally adopted, are 300 pCi/liter or 3.9 x 10^5 MeV α-energy/liter.

Standards are frequently released in the United States from Committees of the Federal Radiation Council. The group will study the aspects of exposure of uranium miners and submit a report to be used as a basis for setting FRC radiation concentration guides for radon and its disintegration products.

8. Raw Water Quality

Raw water quality and standards depend upon the end use. The four main uses are municipal, industrial, agricultural, and recreational (fish and wildlife). Suggested criteria for Oklahoma are presented in Table 8. Water controls the cost of many industrial products. A cheap supply is needed for support of all industry, and requirements for important industries are shown in Table 9.

TABLE 9

TYPICAL INDUSTRIAL WATER REQUIREMENTS

Industrial uses	Rates, gal[a]
Air conditioning	200-23,000/person/season
Brewing	300-1000 bbl beer
Coal washing	600-2400/ton coal
Coke	3600/ton coal
Sugar processing	20,000-25,000/ton sugar
Canning (corn, green beans, peas)	2500-3500/100 cases #2 cans
Meat (packing houses)	550-2000/animal
Poultry	2000/1000 lb live weight
Dairy	340/1000 lb milk and cream
Cheese	200/1000 lb milk and cream
Butter	250/100 lb butter
Vegetable dehydration	500-2000/100 lb
Oil field (secondary recovery)	42,000-504,000/1000 bbl crude
Oil refining	770,000/1000 bbl. crude
Pulp paper	120,000/ton paper
High-grade paper	250,000/ton paper
Textile (cotton)	10,000-40,000/1000 lb goods
Steel	20,000-35,000/ton
Electric power generation	80/kWh

[a] A range indicates variance, depending upon the particular process employed.

REFERENCES

American Standards Association. 1960. "Radiation protection in uranium mines and mills (Concentrators)." American Standard N7.1, p.9

Bureau of Water Resources Research, Oklahoma University. 1962. "Water Quality Criteria for State of Oklahoma." Oklahoma State Dept. of Health.

Busby, C. E. 1961, "Some legal aspects of sedimentation." Proc. Am. Soc. Civil Eng: Hydraulics Div. 4, 151-180.

Canadian Drinking Water Standards and Objectives, 1968 (1969). Prepared by Advisory Committee on Public Health Engineering, Dept. Nat. Health and Welfare, and Canadian Public Health Assoc. Publ. Dept. Natl. Health and Welfare.

Camp, T. R. 1963. Water and its impurities. Reinhold, New York.

Canham, R. A. 1966. "Status of Federal water pollution control legislation." J. Water Pollution Contr. Fed. 38, 1-8.

Clark, J. W. and W. Viessman, Jr. 1965. Water supply and pollution control. International Textbook, Scranton, Penn.

International Atomic Energy Agency. "Basic safety standards for radiation protection (Vienna: IAEA 1962)," p. 46 Table II (Safety Series No. 9).

International Conference on Water Pollution Research 1962, London, Three volumes: Vol. I: Southgate, B. A., Vol. II: Eckenfelder, W. W., Vol. III: Pearson, E. A.; The Macmillan Company - Pergamon Press Ltd., Oxford.

Kneese, A. V. 1964. Economics of regional water quality management. Johns Hopkins University Press, Baltimore, Md.

McKee, J. E. and H. W. Wolf. 1963. Water Quality Criteria. State Water Quality Control Board, California Publ. N. 3-A.

Murphy, E. F. 1961. A study in legal control of natural resources, University of Wisconsin Press, Madison, Wis.

National Academy Sciences, National Research Council 1966. Alternatives in Water Management.

Ontario Water Resources Commission Act. 1960. Revised Statute of Ontario. Frank Fogg, Queen's Printer, Ch. 281, 43 pp.

Radiation Protection. Recommendations of the International Commission on Radiological Protection (ICRP Publ. 2) Rept. Cter. II: Permissible Dose for Internal Radiation. Pergamon, 1962.

Recommendations of the International Commission on Radiological Protection. Br. Radial Suppl. 6 Dec. 1, 1955, 2 p.

Recommended State Legislation and Regulation: (1) Urban Water Supply and Sewerage Systems Act and Regulations, (2) Water Well Constructions, (3) Individual Sewerage Disposal Systems Act and Regulations. U.S. Dept. of Health, Education and Welfare, Publ. Health Service 1965. Supt. Documents, U.S. Govt. Printing Office.

Report International Joint Commission: United States and Canada. 1950. Pollution of Boundary Waters. Washington and Ottawa.

Science Council of Canada. 1968. A major program of Water Resources in Canada, Report 1-5.

Shaw, R. S. and E. R. Segesser. 1958. "State and Interstate Standards for Industrial Wastes." Sewage and Ind. Wastes 30, 909-912.

Taylor, E. F. 1966. "Legal Problems of Water Utilities." J. Am. Water Works Assoc. Sept. 1966, 1205-1216.

U.S. Dept. Health, Education and Welfare, Public Health Service, 1965. Recommended State Legislation and Regulations.

Van Fleet, G. L. 1968. Water Quality Objectives, Senior Course for Water Works Operators, Ontario Water Resources Commission 1-14.

Chapter 2

BIOLOGICAL PARAMETERS
FOR WATER QUALITY

Biological water criteria are as important as chemical criteria. Establishing biological criteria is fraught with difficulties. Wilhm and Dorris (1968) analyzed the problems involved in using biological parameters in water pollution. Chemical substances that affect the quality of water are numerous, they act in a great range of concentrations, and vary continuously and erratically in concentration. Chemical surveys indicate stream conditions only at the times of sampling, and occasional spills of highly concentrated wastes are not easily detected. Thus, sampling procedures become quite important.

The attempt to establish chemical criteria in terms of toxicity to aquatic organisms is difficult and, indeed, may prove to be impossible. The great host of potentially toxic compounds, the vast numbers of species of organisms, the innumerable interaction effects among compounds, and the wide range of effects produced by variations in temperature, dissolved solids, pH, and other physical and chemical factors produce permutations which may exceed

27

the capability of adequate testing. Further, results obtained in the laboratory usually are not transferable to the field, where numerous other environmental conditions may produce unpredictable and unaccountable effects. Even with these problems, biological testing is required.

When wastes are highly treated, chemical testing may not reveal any evident pollutional qualities. However, the receiving stream may be adversely affected by such effluents. Toxic substances too low in concentration to be detectable on a practicable basis may seriously affect populations of aquatic organisms. This requires developing a meaningful approach to the establishment of water quality criteria by the evaluation of biological conditions existing in receiving streams.

I. CONCEPT OF COMMUNITY STRUCTURE

An ecosystem is a natural unit composed of abiotic and biotic elements interacting to produce an exchange of materials. Actions of the abiotic environment and co-actions between biotic components result in a characteristic assemblage of organisms. The complex of individuals belonging to the different species in the ecosystem is referred to as community structure. Natural biotic communities typically are characterized by the presence of a few species with many individuals and many species with a few individuals (Crossley and Bohnsack, 1960). An unfavorable limiting factor such as pollution results in detectable changes in community structure (Mesarovic, 1968; Phillipson, 1966; Brock, 1966).

II. DIVERSITY INDEXES TO ANALYZE COMMUNITY STRUCTURE

The assumption that natural communities represent meaningful assemblages has prompted a diverse series of analyses. One of the simplest and most promising methods of analysis is the diversity index. Diversity indexes are mathematical expressions describing community structure and permitting summarization of large amounts of information about members and kinds of organisms present.

Several diversity indexes have been proposed. One of the first important attempts to interpret animal community structure was that of Fisher et al. (1943). They concluded that the logarithmic series provides an adequate description of the data, and proposed a constant a as an expression of diversity. Preston (1948) stated that the frequency distribution of an animal population is nearly log-normal -- that is, frequency distributions of random samples of ecological assemblages approximate the form of a normal curve on a logarithmic base.

The relationship between number of species and logarithms of the area studied was considered to be linear by Gleason (1922). Margalef (1951) considered that the area studied was proportional to the number of individuals and used this relationship as a measure of community diversity d, where $\bar{d}$ = (s-1)/ln n. Mechinick (1964) used the index d = $s/\sqrt{n}$ to describe community structure of field insects.

After a particular diversity expression is accepted and a meaningful agreement is found between the natural community and a theoretical distribution, a characteristic diversity value can be found to express the structure of each community. In most expressions, maximum diversity exists if each individual belongs to a different species and minimum diversity exists if all individuals belong to the same species. The distribution of individuals among species lies between these extremes in most communities and diversity is intermediate.

An important requisite of a diversity index is independence of sample size. Hairston and Beyers (1954) attempted analysis of populations of soil arthropods with Fisher's logarithmic series and Preston's log-normal series and concluded that both indexes were related to sample size. Mechinick (1964) studied the relationship between species and logarithm of individuals and concluded that this ratio was not a satisfactory measure of diversity because of wide variation with sample size.

A diversity index should reflect not only the distribution of species but should include the relative importance of each species in the community. The structure of the three communities in Table 1 is different; however, the number of individuals, the numbers of species, and the diversity values obtained from (s-1)/ln n or $s/\sqrt{n}$ is the same in all three communities.

TABLE 1

Community Structure of Three Hypothetical
Communities[a]

Community	n_1	n_2	n_3	n_4	n_5	n	s
A	20	20	20	20	20	100	5
B	40	30	15	10	5	100	5
C	96	1	1	1	1	100	5

[a]After Wilhm and Dorris, 1968.

Since considerable variation in biomass exists within and among species, the relative importance of the various species is more adequately expressed in biomass units than in numbers. Differences in diversity values obtained with biomass units and numbers of individuals are shown in Fig. 1. When biomass units are used, the index selected must be dimensionless, otherwise the values generated will depend on the arbitrary choice of weight units. Several of the commonly used diversity indexes do not satisfy the conditions of being independent of sample size, of expressing relative importance of the different species, and of being dimensionless.

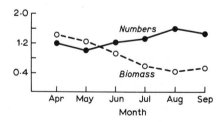

FIG. 1. Differences in diversity of benthic macroinvertebrates in a constant temperature spring obtained with basic data as numbers of individuals and as biomass units. (After Wilhm and Dorris, 1968.)

III. DIVERSITY INDEXES DERIVED FROM INFORMATION THEORY

Margalef (1956) proposed analysis of mixed-species populations by methods derived from information theory. Diversity is equated with the uncertainty that exists concerning the species of an individual selected at random from the population. The more species present in a community and the more equal their abundance, the greater the uncertainty, and hence the greater the diversity. Information content is a measure of uncertainty and thus a reasonable measure of diversity.

The formula given by Brillouin (1960) as a measure of diversity (or information) per individual is

$$\overline{H} = (1/N)(\log N\,! - \sum_{1}^{s} \log N_i)$$

where N is the number of individuals in s species and N_i is the number of individuals in the i-th species. Assuming reasonably large values of N and N_i, the logarithms of the functions may be approximated by Stirling's formula to yield

$$\overline{H} = -\sum (N_i/N)\log_2(N_i/N)$$

The population ratio (N_i/N) is estimated from sample values (n_i/n) to yield

$$\overline{d} = -\sum (n_i/n)\log_2(n_i/n)$$

A theoretical maximum diversity $\overline{d}_{max}$ and a theoretical minimum diversity $\overline{d}_{min}$ can be calculated as follows:

$$\overline{d}_{max} = (1/n)\left[\log_2 n\,! - s\,\log_2(n/s)\,!\right]$$

$$\overline{d}_{min} = (1/n)\left[\log_2 n\,! - \log_2\left[n-(s-1)\right]\,!\right]$$

Then the position of $\overline{d}$ between these extremes can be calculated by the redundancy expression

$$r = \frac{\overline{d}_{max} - \overline{d}}{\overline{d}_{max} - \overline{d}_{min}}$$

Redundancy is an expression of the dominance of one or more species and is inversely proportional to the wealth of species. Thus, indexes derived from information theory permit an expression not only of the compositional richness of a mixed-species aggregation of organisms $\bar{d}$, but also of the dominance of one or more species, r.

The indexes $\bar{d}$ and r possess features that make them reasonable measures of community structure. They are dimensionless equations and numbers or biomass in any units can be used. For example, $\bar{d}$ can be modified to

$$\Sigma - \quad (w_i/w)\log(w_i/w)$$

where w_i is the sample weight of the i-th species, to obtain a dimensionless index based on biomass units. Diversity is then equated with uncertainty regarding biomass rather than numbers. Further, the relative importance of each species in the community is expressed in ratios representing the contribution of each species to total density or biomass. When $\bar{d}$ is applied to the three communities in Table 1, each has a different diversity value. The indexes derived from information theory are also independent of sample size.

Pielou (1966) demonstrated with plant material that as sample size is progressively increased by the addition of new quadrats, the diversity $\bar{d}$ of the pooled sample increases and then levels off. The negligible effect of increased sample size on $\bar{d}$ is shown in Table 2. Successive values in each row represent the values obtained by pooling successive samples from 1 through 10. The value in the first column represents the d of the first sample and the 10th value represents the diversity of all 10 samples pooled. Diversity reached 95% of the asymptotic diversity value by the first sample in the spring, the third sample in the meadow, the fourth sample in the stream, and the fifth sample on a population sampling board. It is unlikely that all species present in the community will be taken in a sampling program. This produces an irregularity in assigning a diversity value with an index such as $(s-1)/\ln n$ or $s/\sqrt{n}$. However, when d is used, the contribution to total diversity will be made by a species that comprises 37% of the sample.

TABLE 2

Diversity values obtained with the index $\bar{d} = -\sum (n_i/n)\log_2(n_i/n)$ by pooling successive samples from 1 through 10 [a]

Habitat	Organisms	No. of samples pooled									
		1	2	3	4	5	6	7	8	9	10
Meadow	Insects	2.37	3.61	3.98	4.02	4.07	4.08	4.16	4.16	4.17	4.16
Stream	Ben. Mac.	2.34	2.98	3.10	3.15	3.16	3.20	3.27	3.30	3.28	3.28
Spring	Ben. Mac.	1.34	1.39	1.40	1.37	1.35	1.38	1.36	1.37	1.38	1.40
Sample board	--	2.18	2.21	2.35	2.63	2.77	2.80	2.87	2.85	2.89	2.86

[a] After Wilhm and Dorris, 1968.

Margalef (1956) collected 1032 individuals of phytoplankton distributed among 21 species from the mouth of the Vigo River. The d of the community was 2.4. Ten of the species were represented by four individuals or less. If these rare species had not been collected, d would have been reduced only to 2.3. If the index $(s-1)/\ln n$ had been used, the diversity of the community would be 2.9 with rare species included and 1.4 if rare species had been overlooked. This demonstrates the importance of estimating ratios rather than numbers of species and individuals (Fig. 2).

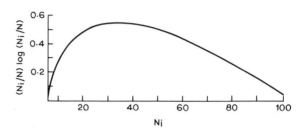

FIG. 2. Relative contribution by the N_i to total diversity, $(N_i/N)\log 2$ (N_i/N), when $N = 100$.

IV. BIOLOGICAL METHODS OF ANALYZING WATER QUALITY

A number of surveys have demonstrated that effluents produce striking changes in the structure of the biotic community. Some species may be unable to survive and others may persist in reduced numbers. With the resulting reduced co-actions, certain species may be able to attain great abundance.

Biotic communities exhibit striking differences in structure at successive stages downstream in a polluted stream. Each successive stage acts on the water in such a way as to change the quality of the water. A distinctive series of populations can be identified until water quality and biotic structure approach the normal situation.

Community structure of benthic macroinvertebrate populations has frequently been used to evaluate conditions in streams receiving organic enrichment. Bottom organisms are particularly suitable for such studies because their relatively low motility does not enable them to escape deleterious substances that enter the environment. Extensive studies were made by

Richardson (1928) on the Illinois River and by Gaufin and Tarzwell (1956) on Lytle Creek. Gaufin and Tarzwell emphasized that associations of benthic macroinvertebrates provide a more reliable criterion of organic enrichment than mere occurrence of indicator species. However, such analyses usually involve long descriptions of associations. Quantitative measures that summarize community structure clearly and briefly are needed in evaluating effects of organic enrichment and other pollution.

V. INFORMATION THEORY METHODS OF ANALYZING WATER QUALITY

The structure of the biotic community can be summarized in diversity indexes derived from information theory. Large numbers of individuals and small numbers of species ordinarily are found in enriched areas of streams receiving organic wastes. Since some species may be superabundant, a large probability exists that an individual observed during sampling belongs to a species previously recognized. Thus considerable repetition of information exists, and redundancy is high. Information per individual is low and is reflected in a low index of diversity. Only a few samples are necessary to describe this low-information system. Stream areas above a sewage outfall and downstream clean-water areas contain smaller numbers of individuals and larger numbers of species. Less repetition of information per individual exists and thus information per individual is greater and redundancy is lower than in enriched areas. More samples are needed to describe this system adequately. Stream areas between the two extremes have intermediate values of diversity and redundancy.

Diversity indexes were used to describe longitudinal variation in community structure of benthic macroinvertebrates in Skeleton Creek, Oklahoma (Wilhm and Dorris, 1966). Municipal and industrial wastes enter the headwaters. Spring and fall values of diversity are shown in Fig. 3. In general, diversity increased progressively downstream. Minimum diversity occurred in the upstream reaches, with maximum $\bar{d}$ at the lowermost station, reflecting the more varied fauna and improved stream conditions. Physicochemical conditions such as chemical oxygen demand, biochemical oxygen demand, dissolved oxygen concentration, turbidity, conductivity, and current velocity

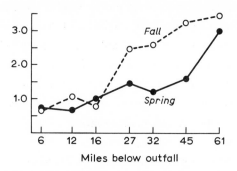

FIG. 3.   Longitudinal variation in d of benthic macroinvertebrates in Skeleton Creek during fall and spring.

in middle-stream reaches resembled conditions downstream in fall and resembled conditions upstream in spring. This pattern was reflected in d, since diversity in middle reaches in fall was similar to that downstream, and diversity in spring was lower and similar to upstream values.

Duncan's multiple range test showed that the five upper stations were not significantly different from each other in mean annual diversity but were different from the two lower stations (Fig. 4). Station 61 was significantly different from all other stations except Station 45. Diversity patterns were similar in fall, except that the three upper stations were significantly different from Stations 27 and 32. In spring, the six upper stations were not significantly different from each other, but were different from the lowermost station.

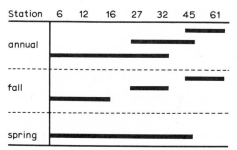

FIG. 4.   Duncan's multiple range test applied to mean annual, fall, and spring $\overline{d}$ in Skeleton Creek. Any two means not underscored by the same line are significantly different. Any two means underscored by the same line are not significantly different.

Species diversity d has been applied to other pollution studies (Table 3). A second diversity analysis of Skeleton Creek was obtained in August, 1967. Harrel (1966) studied community structure of benthic macroinvertebrates in an area of Otter Creek, Oklahoma, that receives oil field brines.

TABLE 3

Examples of Species Diversity d
in Polluted Waters[a]

Area	Pollutants	d		
		Above outfall	Near outfall	Downstream
Skeleton Creek	Domestic, oil refinery	b	0.84	1.59	3.44
Skeleton Creek	Domestic, oil refinery	3.75	0.94	2.43	3.80
Otter Creek	Oil field brines	3.36	1.58		3.84
Refinery Ponds	Oil refinery	b	0.98	2.79	3.17
Keystone Reservoir	Dissolved solids		0.55		3.01
Alamitos Bay	Oil field brines	b	1.49	2.50	b
Alamitos Bay	Oil field brines	b	1.44	2.70	b
Alamitos Bay	Storm sewer	b	1.45	2.81	b

[a]After Wilhm and Dorris, 1968.

[b]Data not available.

Both of these surveys demonstrated a reduction in diversity below an effluent outfall and a progressive increase in diversity downstream. Ewing (1964) studied the structure of insect communities in a series of oil refinery effluent holding ponds. Species diversity in June increased progressively from the pond sampled nearest the effluent to the last pond in the series.

A correlation of -0.93 between conductivity and d have been reported in water. Diversity decreased from 3.01 where mean conductivity was 220 mhos/cm to 0.55 where mean conductivity was 4200.

Subjective evaluation of numbers of species or numbers of individuals can lead to erroneous conclusions. Reich and Winter (1954) studied the benthic fauna of Alamitos Bay, California. According to the authors: "Stations 28 to 30 were located in the vicinity of oil fields. The air at Station 30 was characterized by a petroleum odor. The substratum of these stations possessed a strong sulfied odor, but apparently pollution has not reached toxic levels, as evidenced by the animals present." However, analysis of their data with d demonstrates that oil field brine did affect the biota. Mean species diversity in the vicinity of the oil field was only 1.49 (Table 3). Mean diversity increased to 2.50 in the river downstream. Reich and Winter further reported that "oil fields are located south of Stations 7 to 10, but there was no evidence of pollution in this side channel. Animals were abundant at these stations." Application of d to their data again demonstrated adverse effects of oil pollution. Mean diversity of Stations 7 to 10 was 1.44 and diversity increased to 2.70 in the downstream bay. Diversity was 1.45 in an area that received a storm sewer outfall and increased to 2.81 downstream.

It is apparent that pollution results in a change in the community structure of benthic macroinvertebrates reflected as a depression in d. Values less than 1 have been obtained in areas of heavy pollution, values from 1 to 3 in areas of moderate pollution, and values exceeding 3 in clean water areas. Additional work needs to be done to learn how different types and degrees of pollution are expressed in d.

Application of this method of analysis is relatively simple. Benthic macroinvertebrate populations are sampled at suitable points above and below pollutional outfalls. The distribution of individuals among species is determined by counting, and diversity d and redundancy r are derived from the counts. Variation of these indexes along the stream then gives a measure of the effect of the effluent on stream water quality as reflected in community structure.

Establishment of water quality criteria by the evaluation of biological conditions existing in receiving streams is also an absolute necessity. Effluents produce striking changes in the structure of the benthic macroinvertebrate community. A distinctive longitudinal series of populations can be identified in a polluted stream until water quality and biotic structure approach the normal situation. The structure of a benthic community can be summarized clearly and briefly in diversity indexes derived from information theory. These indexes express the relative importance of each species, are dimensionless, and are independent of sample size.

REFERENCES

Brillouin, L. 1960. Science and Information Theory, 2nd ed. Academic, New York.

Brock, T. D. 1966. Principles of Microbial Ecology, Prentice-Hall, Englewood Cliffs, New Jersey.

Crossley, D. A., and L. C. Bohnsack, 1960. "Long-term ecological study in the Oak Ridge Area: III. The Oribatid mite fauna in pine litter." Ecology, 41: 628-638.

Federal Water Pollution Control Administration, U. S. Department of Interior, May, 1966.

Ewing, M. S. 1964. "Structure of littoral insect communities in a limiting environment: oil refinery effluent holding ponds." Master's Thesis, Oklahoma State University.

Fisher, R. A., A. S. Corbet and C. B. Williams, 1943. "The relation between the number of species and the number of individuals in a random sample of the animal population." J. Animal Ecol. 12: 42-58.

Gaufin, A. R., and C. M. Tarzwell, 1956. "Aquatic Macroinvertebrate communities as indicators of organic pollution in Lytle Creek." Sewage Ind. Wastes, 28: 906-924.

Gleason, H. A. 1922. "On the relation between species and area." Ecology, 3: 158-162.

Hairston, N. G., and G. W. Beyers, 1954. "The soil arthropods of a field in southern Michigan. A study in community ecology." Contrib. Vert. Biol. Univ. Mich. 64: 1-37.

Harrel, R. W. 1966. "Stream order and community structure of benthic macroinvertebrates and fishes in an intermittent stream." Ph.D. Dissertation, Oklahoma State University.

Margalef, R. 1951. "Diversidad de especies en las comunidades naturales." P. Inst. Biol. Apl. 9: 5-27.

Margalef, R. 1956. "Informacion y diversidad espicifica en las cominudades de organismos." Invest. Pesquera, 3: 99-106.

Mechinick, E. F. 1964. "A comparison of some species-individuals diversity indices applied to samples of field insects." Ecology, 45: 859-861.

Mesarovic, M. D. 1968. Systems Theory and Biology. Springer - Verlag, New York.

Phillipson, J. 1966. Ecological Energetics. Edward Arnold, London.

Pielou, E. C. 1966. "The measurement of diversity in different types of biological collections." J. Theor. Biol., 13:

Preston, F. W. 1948. "The Commonness, and rarity of species." Ecology, 29: 254-283.

Reisch, D. J., and H. A. Winter, 1954. "The ecology of Alamitos Bay, California, with special reference to pollution." Calif. Fish Game, 40: 105-121.

Richardson, R. E. 1928. "The bottom fauna of the middle Illinois River, 1913-1925." Bull. Illinois Nat. Hist. Surv. 17: 387-473.

Wilhm, J. L. and T. C. Dorris, 1966. "Species diversity of benthic macroinvertebrates in a stream receiving domestic and oil refinery effluents." Am. Midland Naturalist, 76: 427-449.

Wilhm, J. L. and T. C. Dorris, 1968. "Biological parameters for water quality criteria." BioScience, 18: 477-481.

Chapter 3

UNITS, ABBREVIATIONS, AND TERMINOLOGY

An important part of a wastewater treatment is the laboratory. The analyses performed there determine the efficiency of the operation and corrections to be made. Only the most essential tests and unit operations are discussed herein. Important but not all details of analyses are included. The complete methods will be found in Standard Methods for the Examination of Water and Wastewater (1965), published by the American Public Health Association. The book may be purchased from the APHA, the Water Pollution Control Federation, or the American Water Works Association.

The metric system is used in all laboratory work. However, the gallon is still preferred by many engineers and thus conversion is required. In comparing liters and quarts, a liter has 1000 ml and a quart 946 ml. Thus, 1.057 quarts equal 1 liter, or 1 quart = 0.946 liter.

I. CONVERSION FACTORS

The units most frequently used are grams (g), milligrams (mg), liters (liters), milliliters (ml), part per million (ppm), or milligrams per liter (mg/liter). The metric system is much easier to use:

1 liter	=	1000 ml
1 gram	=	1000 mg
1 mg	=	0.001 g
1 mg	=	1000 microgram (μg)
1 microgram	=	0.001 mg
1 ug	=	1000 picograms (pg)
1 pg	=	1 micromicrogram ($\mu\mu$g)

43

Parts per million (ppm) or milligrams/liter (mg/liter) is a weight ratio. Any unit may be used; pounds per million pounds, grams per million grams; gallons per million gallons, etc.

1 ppm = 8.34 lb/million gallons, or 1 lb/million pounds, or 1
milligram/liter (mg/liter)

The most frequently used conversion factors in wastewater treatment are shown in Table 1.

TABLE 1

CONVERSION FACTORS ENGLISH TO METRIC SYSTEM

1 To convert from	2 To	3 Multiply by[a]	To convert Column 2 to Column 1 Multiply by
acre-feet	cubic meters	1233	8.11×10^{-4}
cubic feet (cu ft) (U.S.)	cubic centimeters	28,317	3.53×10^{-5}
cubic feet (cu ft) (U.S.)	cubic meters	0.0283	35.3
cubic feet (cu ft) (U.S.)	liters	28.3	0.035
cubic feet/minute	cubic centimeters/ second	472	0.0021
cubic feet/minute	liters/second	0.472	2.119
cubic feet/second	liters/minute	1699	5.886×10^{-4}
cubic inches (U.S.)	cubic meters	1.64×10^{-5}	61,024
cubic inches (U.S.)	liters	0.0164	61,024
cubic inches (U.S.)	milliliters (ml)	16.387	0.0610
feet (U.S.)	meters	0.3048	3.281
feet (U.S.)	millimeters (mm)	304.8	3.28×10^{-3}
feet/minute	centimeters/second	0.508	1.97
feet/minute	kilometers/hour	1.097	54.68
feet/minute	meters/minute	18.29	0.0547
feet/second^2	kilometer/hour/ second	1.0973	0.911
gallons (U.S.)	cubic centimeter (ml)	3785	2.64×10^{-4}
gallons (U.S.)	liters	3.785	0.264
gallons/minute	liters/second	0.063	15.85
grains	grams	0.0648	15.432
grains	milligrams (mg)	64.8	0.1543
grains/gallon (U.S.)	grams	0.0171	58.417
grains/gallon (U.S.)	ppm	17.1	0.0584
inches (U.S.)	centimeters (cm)	2.54	0.3937
inches (U.S.)	millimeters (mm)	25.4	0.0394
miles (U.S.)	kilometers (km)	1.609	0.6214
miles (U.S.)	meters	1609	6.214×10^{-4}
miles/hour	centimeters/second	44.7	0.0224
miles/hour	meters/minute	26.8	0.0373
miles/minute	kilometers/hour	96.6	1.03×10^{-3}

TABLE 1

CONVERSION FACTORS ENGLISH TO METRIC SYSTEM (Continued)

1 To convert from	2 To	3 Multiply by[a]	To convert Column 2 to Column 1 Multiply by
ounces (avoirdupois)	grams	28.35	0.3527
ounces (U.S. fluid)	millimeters	29.6	0.0338
ounces (U.S. fluid)	liters	0.0296	33.81
pounds (av)	grams	453.6	0.0022
pounds (av)	kilograms	0.4536	2.205
pounds/cubic foot	grams/liter	16.02	0.0624
pounds/foot	grams/centimeter	14.88	0.067
pounds/gallon (U.S.)	grams/millimeter	0.12	8.345
pounds/gallon (U.S.)	grams/liter	119.8	8.34×10^{-3}
quart (U.S. liq)	milliliter	946	0.0021
quart (U.S. liq)	liters	0.946	2.113
square feet (U.S.)	square centimeters	929	1.08×10^{-3}
square feet (U.S.)	square meters	0.929	10.76
square inches (U.S.)	square centimeters	6.45	0.115

[a]To convert from the metric system to the English system, take the reciprocal of the figures in the third column.

For example: To convert from acre-feet to cubic meters, multiply by 1233, but to convert from cubic meters to acre-feet multiply cubic meters by $\frac{1}{1233}$ or 8.11×10^{-4}. Common abbreviations used in water pollution are presented in Table 2.

TABLE 2

ABBREVIATIONS FOR UNITS OF MEASUREMENT

cfm	=	cubic feet per minute
cfs	=	cubic feet per second
cc	=	cubic centimeters
cm	=	centimeter(s)
ft	=	foot or feet
ft^2	=	square feet
ft^3	=	cubic feet
fps	=	feet per second
gal	=	gallon(s)
gpd	=	gallons per day
gpm	=	gallons per minute

TABLE 2

ABBREVIATIONS FOR UNITS OF MEASUREMENT (Continued)

g	=	gram(s)
hr	=	hour(s)
in	=	inch(es)
in.^{2}	=	square inch(es)
in.^{3}	=	cubic inch(es)
lb	=	pound(s)
km	=	kilometer(s)
m	=	meter(s)
mg	=	milligram(s)
mg/liter	=	milligram(s) per liter
mil gal	=	million gallons
mgd	=	million gallons per day
min	=	minute(s)
ml	=	milliliter(s)
MLSS	=	mixed liquor suspended solids
oz	=	ounce(s)
ppm	=	parts per million = mg/liter
sec	=	second(s)
ss	=	suspended solids
VSS	=	volatile suspended solids
yd	=	yard(s)
SVI	=	sludge volume index

II. SAMPLING

In collecting samples -- whether from a sewage plant or industrial waste plant -- it is absolutely necessary to composite samples in proportion to flow (Table 3).

TABLE 3

CHART FOR COMPOSITE SAMPLING

Time	Flow mgd	Factor	Volume of portion in ml
6:00 AM	0.3	100	30
7:00 AM	0.6	100	60
8:00 AM	0.8	100	80
9:00 AM	1.0	100	100
10:00 AM	1.5	100	150
11:00 AM	1.4	100	140
12:00 Noon	1.2	100	120
1:00 PM	1.0	100	100
2:00 PM	1.0	100	100
3:00 PM	0.9	100	90
4:00 PM	0.8	100	80
5:00 PM	0.8	100	80
			1130 ml

For this example a composite sample of 1130 ml was collected. If a gallon sample were needed, the factor would be tripled to 300, making each portion three times as big and yielding 3390 ml of sample. An analysis is no better than the sample analyzed and cannot be accurate if the sample is not representative. Equipment cannot be sized without proper analysis.

The more important terms encountered in water pollution are compiled below.

ACIDITY: Ordinarily expressed as pH below 7.

ACTINOMYCETES: A group of branching filamentous bacteria repro-
                      ducing by terminal spores. They are common in
                      the soil. These bacteria are known to cause offen-
                      sive taste and odor in water.

ALGAE: An assemblage of nonvascular plants consisting of
                      a great number of genera and species, both micro-
                      scopic and large in size. Algae carry on photosyn-
                      thesis, differentiating them from fungi.

ALGICIDE: Any substance that kills algae.

ALKALINITY: Ordinarily expressed as pH above 7.

ANTAGONIST: An agent, such as a chemical, that tends to nullify
                       the action or effect of another agent or force when
                       both are present concurrently in a medium.

AQUAMARSH: A stage in the evolution between open water and
                       land marsh.

AQUATIC   The aquatic plants of lakes and other kinds of
   PLANTS: standing water bodies are those whose seeds ger-
                       minate in the water or in the lake-bottom soil.

AQUIFER: A layer of rock, sand, or gravel through which
                       water can pass.

AREAL   An area of $400\mu^2$ on a microscopic slide or count-
   STANDARD UNIT: ing chamber, used as a unit in designating the con-
                       centration of microorganisms, such as algae, in
                       water. One micron = 1×10^{-3} ml.

ARTESIAN: Describes underground water trapped under pres-
                       sure between layers of impermeable rock. An
                       artesian well is one that taps artesian water.

BACTERIA: One-celled microorganisms.

BANK: Pertaining to a lake, the sharply rising ground or
                       abrupt slope, usually wave-cut and presenting a
                       nearly vertical front.

BASIN: An impoundment.

BEACH: The width of the shore zone lapped by waves.

BED: Submerged land surface of a lake basin.

BENTHOS: A limnological term for a whole group of bottom
                       dwelling organisms of a lake.

BIOCHEMICAL   The amount of dissolved oxygen in parts per million
   OXYGEN   required by organisms for the aerobic biochemical
   DEMAND: decomposition of organic matter present in water.

BIOCIDE:	An organic chemical agent used for the destruction of living organisms such as pests, nuisance organisms, and weeds, aquatic or terrestrial (e.g. pesticide, algicide, weedicide, fungicide).
BIODEGRADABLE ORGANICS:	Those that can be broken down by microorganisms to form stable compounds such as CO_2 and water.
BOG:	Any wetland feature or body of land, characterized by a spongy, miry surface.
BORISM:	Poisoning by a boron compound.
BREAKWATER:	A structure for breaking the force of waves.
CAE:	Carbon Alcohol Extractibles: substances extracted by alcohol, from an activated carbon column ("filter") through which unchlorinated water has been passed.
CANAL:	A dredged canal connecting two bodies of water.
CAPE:	A rounded projection out into the water.
CAPILLARITY:	The force that causes water to rise in a constricted space through molecular attraction, often against the pull of gravity.
CAT ICE:	Ice forming a thin shell from under which the water has receded.
CATCHMENT BASIN:	The entire area from which drainage is received.
CATHARSIS:	Purgation of the alimentary canal.
CAUSEWAY:	A raised way or road made across wet or marshy ground, or across the surface of a lake.
CCE:	Carbon Chloroform Extractibles: substances extracted by chloroform, from an activated carbon column ("filter") through which unchlorinated water has been passed.

CESSPOOL: Structure designed to hold sewage from a residence.

CHLORINATION: The application of a chlorine compound to water for
                           the purpose of disinfection, and frequently also for
                           accomplishing other biological or chemical results
                           in the treatment of water.

CHOLINERGIC: Stimulated, activated, or transmitted by acetyl-
                           choline. The term is applied to those nerve fibres
                           that liberate acetylcholine at a synapse when a
                           nerve impulse passes, i.e., parasympathetic nerve
                           endings.

CHOLINESTERASE: An esterase (enzyme) present in all body tissues
                           which hydrolyzes acetylcholine into choline and
                           acetic acid.

COAGULATION: A water treatment process comprising a series of
                           chemical and mechanical operations involving the
                           addition of chemicals (coagulants) to water, their
                           mixing and uniform distribution through the water,
                           and the building up of a readily settleable floc,
                           usually by prolonged agitation of the water. This
                           process may be distinguished into two separate
                           phases, "mixing" and "flocculation."

COAST: Shorelands of the Great Lakes, or land bordering
                           seas.

COLIFORM   The "coli aerogenes" group. A test for the pres-
BACTERIA: ence of coliform bacteria is commonly used to
                           determine the presence of fecal contamination.

COMPLETE   A system of water treatment processes including
TREATMENT: coagulation (mixing and flocculation), sedimentation
                           (settling or clarification), filtration, and disinfec-
                           tion or, as used in this document, a combination or
                           modification of these processes but always including
                           disinfection.

CONDENSATION:	The transformation of water from a vapor to a liquid.
CONE OF DEPRESSION:	A conical dimple in the water table surrounding a well, caused by pumping.
CONSUMPTIVE USE:	The use of water, especially in irrigation, in such a way that it is converted to vapor and returned to the atmosphere.
CONTAMINATION:	The introduction into otherwise potable water of toxic materials, pathogenic or undesirable micro-organisms, and other deleterious substances that make the water unfit for use; a specific type of "pollution."
CONTROL AGENCY:	A governmental health or water quality and pollution control agency vested with the responsibility of assessing the safety and approving a drinking water supply.
COOLING WATER:	Water that has been used primarily for cooling in an industrial manufacturing process.
COROLLARY POLLUTANT:	Any chemical substance or biological organism that is indigeneous to the waterbody (e.g., naturally occurring chemicals, algae, fungi, invertebrate larvae, etc.).
COUPON INSERTION (COUPON TEST METHOD):	A method for the quantitative measurement of corrosion and scaling characteristics of water on metal coupons in the absence of heat transfer. Carefully prepared metal coupons are installed in contact with flowing water for a measured length of time and, after removal, the coupons are examined for depth, distribution, weight, and character of foreign matter on coupons, and the corrosion characteristics of the water by the difference in weight.

CURRENTS: Movements or flows of the water.

CYANOSIS: Blueness of the skin and the white of the eye, due
                     to insufficient oxygenation of the blood resulting in
                     the formation of methemoglobin.

DAM: A structure designed to hold back a flow of water.

DAM LAKE: One created by a dam.

DAM POND: An impoundment behind a man-made dam.

DEAD LAKE: Colloquialism for lakes that have become filled
                     with vegetation.

DEADHEAD: A log lying on the bottom of a lake or in the bed of
                     a river.

DEADWATER: A stream of water that appears to have no flow.

DEPTH: As applied to water, generally implies the whole
                     vertical distance between the surface and the
                     bottom at a given position.

DESALINATION: Process of removing salt from water.

DESICCATION: Loss of water by direct evaporation.

DESTRATIFICATION: Artificial circulation or mixing of the water.

DIATOMS: Microscopic plants that occur abundantly as float-
                     ing forms in plankton.

DIGESTION: The dissolution of insoluble and soluble solids to
                     CO_2, H_2O, etc. by microbes.

DIKE: Artificial embankment.

DISCHARGE: The rate of flow of surface or underground water.

DISINFECTION: A process of destroying microorganisms and vege-
                     tative cells occurring in water by the application of
                     a chemical agent (disinfectant) such as a chlorine
                     compound (see Chlorination).

DISPOSAL AREA:	Area of land, usually in a marsh or swamp, where dredged material from a lake improvement project is deposited.
DISSOLVED OXYGEN:	The amount of dissolved oxygen in parts per million present in water.
DIURESIS:	Increased secretion of urine.
DIVERSION:	Draining, pumping, siphoning, or removal of water from a lake in any manner that is not natural.
DRAG LINE:	Machine sometimes used to remove bottom material from a lake.
DRAIN:	A surface ditch or underground pipe for the purpose of conducting fluids.
DRAINAGE BASIN:	The geographical area within which all surface water tends to flow into a single river or stream via its tributaries.
DRAWDOWN:	Term for a decrease in the levels of reservoirs or other impoundments.
DRIFT:	Speed of a current.
DROP-OFF:	Scarp or bank of a subaqueous terrace or littoral shelf.
EFFLUENT:	An outflowing surface stream.
EMESIS (EMESIA):	Vomiting or an act of vomiting.
ENRICHED LAKE:	One that has received inputs of nitrates, phosphates, and other nutrients, thereby greatly increasing the growth potential for algae and other aquatic plants.
ENTERIC:	Pertaining to the intestines.
ENTEROVIRUS:	Virus of intestinal origin. As used in this document, this term refers to a virus discharged from the intestines of warm-blooded animals including man.
EPHEMERAL LAKES:	Short-lived lakes and ponds.

EPILIMNION: In a thermally stratified lake, the turbulent layer
                       of water that extends from the surface to the
                       thermocline.

EROSION: The wearing down of the earth's surface by water.

EUPHOTIC   Depth zone through which light penetrates water.
ZONE: Effective in photosynthesis.

EUTROPHIC   "Rich" lakes. Those well provided with the basic
LAKES: nutrients required for plant and animal production.

EVAPORATION: Transformation of water into vapor.

EVAPOTRANS-   The process by which water, evaporated from the
PIRATION: earth and given off by plants and animals, is re-
                       turned to the atmosphere as vapor.

FECAL COLIFORM   A member of the group of coliform organisms, and
ORGANISM: of fecal origin (i.e., from the intestines of warm-
                       blooded animals including man).

FALSE BEACH: A bar, above water level, a short distance off
                       shore.

FARM POND: A small shallow structure for the impoundment of
                       water.

FAST ICE: Ice attached to and extending out from the land.

FAULT: A break in the earth's crust.

FECAL MATTER: Human and animal excrement.

FETCH: On a lake surface, the reach, or longest distance
                       over which the wind can sweep unobstructed.

FINGER LAKES: Long, narrow lakes occupying deep troughs in
                       deeply eroded, straight pre-glacial valleys in
                       glaciated regions.

FISH KILL: Destruction of fish in lakes or ponds due to pro-
                       longed ice and snow cover resulting in oxygen
                       deficiency.

FISSION LAKES:	Lakes that represent division of or separation from an original single body of water.
FLOCCULATION:	A phase of the coagulation process, involving agitation of the water at lower velocities than for "mixing," for a measured length of time during which the very small particles grow, coalesce, and agglomerate into well-defined hydrated flocs of sufficient size to settle readily (see Coagulation).
FLOODPLAIN:	A strip of flat land bordering a stream or river consisting of sediment laid down over the centuries by the river.
FLOODING:	Water bodies that inundate or cover flat lands as a thin sheet.
FLOWAGE:	Volume or dimensions of the water of a stream.
FLUORIDATION:	The controlled adjustment of the fluoride ion concentration of drinking water to a selected concentration.
FLUSHING PERIOD:	Time required for an amount of water equal to the volume of the lake to pass through its outlet.
FLUVIATILE LAKE:	Lake formed in the flood plain of a river such as an ox-bow lake, or other water body formed as a result of stream erosion and deposition.
FOREBAY:	Small pond or reservoir at the head of the penstock of a power dam.
FORESHORE:	Part of a shore or beach normally subject to the uprush and backrush of waves.
FOSSIL LAKE:	One that has been extinct for a long period of time.
FRESHET:	Small affluent streams having a high rate of flow.
FUNGUS: (pl. FUNGI)	A group of organisms without chlorophyll or other photo- or chemosynthetic pigments, obtaining nutrients from preformed organic matter by absorption (i.e., "heterotrophic"); unicellular

or filamentous organisms. Some are parasitic in living tissues and may produce pathogenic conditions; others may cause odor and taste in water, and may cause nuisance conditions in sewers and watercourses.

GABION: Specially designed basket or box of corrosion resistant wire designed to hold rocks to form groins, sea walls, etc.

GLACIAL LAKES: Lakes formed as a result of glacial action.

GROIN: Long, narrow wall-like structure extending out into a lake normal to the shore.

GROUNDWATER: All subsurface water.

HEADRACE: A race constructed to lead water to a water wheel or into an industrial building.

HEADWATER: Beginning of a stream or river; its source.

HEALTH: A state of complete physical, mental, and social well-being and not merely the absence of disease or infirmity; it includes satisfaction of the esthetic aspirations of man.

HEALTH AGENCY: A governmental department of health or a health unit.

HOMOTHERMOUS: Maintaining a uniform bodily temperature independent of the environmental temperature; birds and mammals (including man) are homothermous animals.

HYDROGRAPH: A graph showing the stages or variations in water level over a period of time.

HYDROGRAPHIC BASIN: Area of the watershed of a lake plus the area of the lake.

HYDROLOGIC CYCLE: The process by which water constantly circulates from the sea to the atmosphere to the earth and back to the sea again.

HYDROLOGY:	The scientific study of water found on the earth's surface, in its subsurfaces, and in the atmosphere.
HYDROLYSIS:	The process by which a compound reacts chemically with water and forms a new substance.
HYPOLIMNION:	The water below the thermocline.
IAC:	Intermediate-Aerogenes-Cloacae group of bacteria; may be found in fecal discharges, but usually in smaller numbers than Escherichia coli. Also commonly present in soil and in water polluted sometime in the past; survive longer in water than do fecal coliform organisms; tend to be somewhat more resistant to chlorination than E. coli or the commonly occurring bacterial intestinal pathogens.
ICRP:	International Commission on Radiological Protection.
INFLUENT:	An in-flowing stream.
INFILTRATION:	The method by which surface water is soaked into the ground through tiny openings in the soil.
ISLAND:	An area of land within a lake.
JETTY:	Structure similar to a groin.
LAGOON:	Water body in depression.
LEACHING:	Process by which water, seeping through earth and rocks, dissolves and carries away certain minerals or compounds.
LIMNETIC ZONE:	In lakes partly occupied by emergent vegetation, the area of open water.
LIMNOLOGY:	Science that deals with lakes, and by extension with all inland waters. It is concerned especially with the biology of the waters and bottoms.
MARGIN:	Land immediately bordering the water line.

MBAS:	Methylene Blue Active Substances. Anionic surface-active agents (synthetic detergents) that form a blue-colored salt upon reaction with methylene blue.
MESOTROPIC LAKE:	One that is intermediate in fertility; neither notably high nor notably low in its total productivity. Intermediate between oligotrophic and eutrophic.
METABOLIC PROCESS:	The "building up" (anabolic) and the "tearing down" (catabolic) biochemical processes of living cells.
METABOLITE:	Any substance produced by a metabolic process of living cells.
METHEMO-GLOBINEMIA:	A disease caused by the presence of methemoglobin in the blood. Methemoglobin is the modified oxy-haemoglobin or an oxidized heme containing ferric iron combined with the normal globin.
MF METHOD:	Membrane Filter Method.
MICRON:	A unit of linear measurement, often used for describing the dimensions of a microorganism or microscopic particle. One micron is the equivalent of one one-thousandth of a millimeter (1×10^{-3} mm or 0.93937×10^{-3} in.).
MICROORGANISM:	Any minute plant or animal organism invisible or barely visible to the unaided eye (e.g., bacteria, viruses, algae, etc.).
MPC_w:	Maximum Permissible Concentration in water.
MPN:	Most probable number (of coliform bacteria), method of quantitative analysis of coliform organisms (Multiple tube dilution analysis).
NARROWS:	Constriction in the width of a lake.
NOT DETECTABLE:	Below limit of detectability by a specified method of analysis.
O.D.:	Abbreviation for oxygen demand.

ORGANIC NITROGEN:	The nitrogen combined in organic molecules such as proteins, amines, and amino acids.
ORP (OXIDATION = REDUCTION POTENTIAL):	Indicates the degree of completion of the chemical reaction by detecting the ratio of ions in the reduced form to those in the oxidized form as variations in electrical potential from an ORP electrode assembly.
OUTFALL:	Term used for the end of a pipe, ditch drain, or distributor that carries an effluent into a lake or stream.
OXBOW:	A curved lake, created when a bend is abandoned by a river that has changed its course.
OXYGEN DEMAND:	Total oxygen consumed by the chemical oxidation of organic material in water.
PATHOGENIC: (PATHOGENETIC)	Giving origin to or resulting in disease or morbid symptoms.
PATHOGENIC BACTERIA:	Bacteria that cause disease.
PENSTOCK:	Closed tube used to conduct water under pressure from a reservoir to a turbine house.
PERMEABILITY:	Capacity of a solid to allow the passage of a liquid.
PHOTOSYNTHESIS:	A biochemical process by which an organism manufactures sugar or other carbohydrates from inorganic raw materials with the aid of light and pigments such as chlorophyll. Organisms possessing the ability to carry on photosynthesis are termed "photosynthetic" organisms.
PICOCURIE:	A synonym for micromicrocurie; a subdivision of the unit curie which expresses the rate of disintegration of a radionuclide, in which 3.7×10^{10} atoms disintegrate per second. One picocurie or micromicrocurie is the equivalent of 1×10^{-12} Ci.

PIEZOMETRIC SURFACE:
The theoretical level to which water should rise under its own pressure if tapped by well or spring.

PILING:
Usually wooden posts driven into the lake bottom to support docks or other structures.

PLANKTON:
A term for an assemblage of microorganisms, both plant and animal, that float, drift, or swim in the water.

POLLUTANT:
A substance, biological, inorganic, organic, or radioactive, that brings about a condition of pollution.

POLLUTION: (Water)
Anything causing or inducing objectionable conditions in any watercourse and affecting adversely the environment (ecology) and use or uses to which the water thereof may be put.

POND:
Generally applied to small impoundments for a source for livestock and for other uses on farms. However, the term can be used for large impoundments of water.

POROSITY:
The ability of rock and other earth materials to hold water in open spaces or pores.

POTABLE WATER:
Water that is drinkable and usable for culinary purposes, as a result of being free of pathogenic organisms (or their indicators), toxic substances, objectionable taste, odor, and color, and other undesirable physical, chemical, and biological characteristics.

ppb:
Abbreviation for parts per billion.

ppm:
Abbreviation for parts per million.

PRECIPITATION:
The discharge of condensed water vapor by the atmosphere in the form of rain, hail, sleet, or snow.

PRIMARY POLLUTANT:	Any chemical substance or biological organism that is added directly or indirectly to water as a result of human activities within the watershed (e.g., toxic chemicals, enteric bacteria, enteroviruses, etc.).
PUTRESCIBLE WASTE:	Matter, usually of animal origin, that is subject to the process of putrefaction.
RACE:	An artificial channel, or canal leading from a dam to a water wheel, or other contrivance for generating power.
RAW WATER:	Surface or underground water available as a source of drinking water supply but which has not been treated or "purified."
REEF:	A shoal consisting of a ridge of sand, gravel, or hard rock.
RESERVOIR:	Applied to waters held in storage in either artificial or natural basins.
REVETMENT:	A facing of stone, concrete, or other material to protect the banks of a lake from wave erosion.
RICH LAKES:	A term applied to those "rich" in nutrients and capable of supporting an abundant flora and fauna.
RIPARIAN:	A person with rights to water by virtue of ownership of land bordering the bank of a stream or waterline of a lake.
RIPPLE:	A very small wave. One whose period is arbitrarily defined as three seconds or less.
RIPRAP:	Coarse stones, natural boulders, or rock fragments laid against the basal slope of a bank for the prevention of wave cutting.
ROUGH FISH:	Those species such as carp and sucker, considered undesirable, predatory, or obnoxious by anglers.

SATELLITE LAKES:	One or more small lakes disconnected and separated from but associated with a single large lake in a single basin.
SCOUR:	Lakes occupying depressions, or basins made by gouging and abrading actions of glaciers passing over soft rocks or moving in pre-existing valleys.
SEDIMENT:	Tiny particles of rock and dirt carried by water, which eventually settle to the bottom.
SEEPAGE LAKE:	One that loses water mainly by seepage through the walls and floor of its basin.
SENESCENT LAKE:	One nearing extinction; especially from filling by the remains of aquatic vegetation.
SETTLEABLE SOLIDS:	Those in suspension that will pass through a $2000\text{-}\mu$ sieve and settle in 1 hr under the influence of gravity.
SETTLING BASIN:	An artificial basin for collecting the sediment of a river before it flows into a reservoir.
SEWAGE:	An inclusive term applied to all effluent carried by a sewerage system.
SHORE:	Technically, the zone of wave action on land.
SILTING:	The filling of reservoirs by sediments.
SKIM ICE:	When freezing first commences on a lake the first ice crystals formed are free floating or weakly attached.
SKIN ICE:	The first film or crust of newly formed ice.
SLIME:	Soft fine oozy mud.
SLOUGH:	Standing water bodies.
SLUDGE:	Term used to define settled waste material (in liquid or solid form) in sewage treatment plants.
SLUDGE DRYING BED:	Used to dewater pretreated sludge and reduce volume to be handled for disposal.

SNAG LAKE:	A lake containing trunks and branches of trees, or the pointed roots of stumps lying on the bottom.
SPRING:	An opening in the surface of the earth from which groundwater flows.
STOCKED LAKE:	A lake planted with fish of a desirable species.
STORAGE RESERVOIR:	In a storage reservoir water is impounded back of a high dam and held for later use.
STRAND:	The strip at the base of a shore cliff, or along a shore, that is lapped by waves.
STRATIFIED FLOW:	When a difference exists between the density of the inflowing water and that of a lake or reservoir.
STRATIFIED LAKES:	In deeper lakes, especially in temperate regions, the water from top to bottom exhibits differences in temperatures.
SUBAQUATIC PLANTS:	Emergent plants; or hydrophytes that are not submersed.
SURF:	The effect produced by the break of a wave as it enters shallow water or a shallow shore zone.
SURFACE WATERS:	That lying on the surface of the land in contrast to underground water.
SUSPENDED SOLIDS:	Those that can be removed by filtration or centrifugation.
SWELL:	A wave that continues after the wind has ceased.
SWIMMERS ITCH:	A rash produced by a parasitic flat worm (in the cercarial stage of its life) that penetrates the skin of bathers.
TAILING PONDS:	Enclosures, or basins, constructed for the disposal of mine tailings.
TERRACE:	A plateau on the side of a valley, representing an old floodplain no longer reached by the river below.

THERMOCLINE:	In thermally stratified lakes, the layer below the epilimnion.
TOTAL SOLIDS:	The sum of suspended solids and dissolved solids.
TOXIC:	Pertaining to, due to, or of the nature of, a poison.
TOXICITY:	The quality of being poisonous, especially the degree of virulence of a poison.
TRANSPIRATION LOSSES:	Water consumed by emergent and floating lake plants, voided as gas through specialized leaf cells.
TRUE COLOR:	Color attributable to substances in solution after the suspensoids have been removed by centrifugation (not by filtration).
TRUE COLOR UNIT:	A unit of color equivalent to the color produced by 1.0 mg/liter of chloroplatinate ion by the Platinum-Cobalt Method.
TURBIDITY:	Cloudiness caused by sediment suspended in water.
UNDERFLOW:	The downstream movement of groundwater through permeable rock beneath a riverbed.
VIRUS:	An extremely small living organism or nonliving particle; it is sometimes found in lake water.
WAKE:	The track left in the water by a moving boat.
WASTE STABILIZATION POND:	A shallow artificial pond constructed for the stabilization of minicipal and industrial wastes.
WATER BLOOM:	A prolific growth of plankton.
WATER QUALITY:	The graded value of a single property, or the characteristics as a whole, in relation to a particular use.
WATERLINE:	The line of contact between the still water of a lake or pond and the bordering land.
WATERSHED:	The whole surface drainage area that contributes water to a lake.

WATER TABLE: The level to which groundwater rises, or the surface
                   of the zone of saturation.

WATERWAY: A navigable body of water, natural or artificial, that
                   serves as a water highway or water road.

WAVE: An undulation or ridge on a water surface.

WILD RICE: An aquatic grass, fairly common in northern lakes.

YIELD: The quantity of water that can be taken, continuously,
                   for any particular economic use.

ZOOPLANKTON: Animal microorganisms living unattached in the
                   water.

Waste analyses are needed and required for all water water. These analyses guide the engineer in taking corrective measures. A typical report analysis is shown in Table 4.

Pollutants fall into three basic categories. They can be further classified by unwanted properties and chemical or biological origins.

1. Floating materials: oils, greases, foam, and other solids that are lighter than water.
2. Suspended matter: insoluble materials such as mineral tailings.
3. Dissolved impurities: acids, alkalies, heavy metals, insecticides, cyanides, and other toxics.

Subcategories affecting these three are temperature, color, taste and odor, and radioactivity.

1. Temperature: temperature controls all forms of aquatic life and the solubilization and deposition of inorganic materials.
2. Color: color does interfere with biological activity by retarding transmission of sunlight into the stream. Color also indicates the presence of undesirable dissolved and suspended solids. For example, dissolved chromate ions are yellow; copper is blue.
3. Taste and odor: taste and odor are important indicators of undesirable impurities.

TABLE 4

Report of Wastewater Analysis

Source: typical municipal waste-water			
Date collected -- May 8, 1967			mg/l
		Iron	0.04
Date analyzed -- May 9, 1967		Copper	0.1
		Manganese	0.0
pH value	6.9	Aluminum	0.05
Alkalinity to pH 8.2 as $CaCO_3$	0.0 mg/liter	Cadmium	0.0
Alkalinity to pH 4.6 as $CaCO_3$	260 mg/liter	Nickel	0.0
		Lead	0.0
Total acidity as $CaCO_3$	0.0 mg/liter	Zinc	0.02
FMA as $CaCO_3$	0.0 mg/liter		
Conductivity (μ mhos/cm)	--		
Chloride	110	Syndet-- anionic	5.1
Sulfate	102	Syndet-- nonionic	1.1
Silica	27	Color --APHA units	--
Nitrate	0.0	Jackson turbidity	
Phosphate--total	30.0	units	--
Phosphate--ortho	10.0	Hex-chromium as CrO_4	0.0
Hardness as $CaCO_3$	168	Tri-chromium as CrO_4	0.0
Calcium	51		
Magnesium	10	Ether soluble organic	52
Dissolved solids	200	Solvent soluble	
Suspended solids	300	organic	72
Volatile suspended solids	250	Total organic carbon	120
Total solids	500	Chemical oxygen	
Volatile total solids	350	demand	350
Settleable solids	ml/liter	Dissolved oxygen	0.0
15 min	5	Biochemical oxygen	
30 min	7	demand	200
60 min	8	Cyanide	0.01
90 min	9	Phenol	0.2
120 min	10	Fluoride	0.4
		Total nitrogen	50
		Organic nitrogen	20
		Ammonia nitrogen	30
		Albuminoid nitrogen	--

4. Biological matter: fungi are similar to bacteria. Algae in excess amount can function as heterotrophs and use up oxygen and bring taste and odor problems. Microscopic animals such as protozoa are valuable as bacteria scavengers.

A wide variety of organic and inorganic chemical compounds find their way into ground and surface waters. All are potential pollutants.

Acids	Fluorides	Oxidizing agents
Alkalies & hydroxides	Fungicides	Phenols
	Hardness	Potassium
Arsenic	Hydrocarbons	Reducing agents
Barium, Boron	Hydrogen sulfide	Selenium
Cadmium, Cesium	Insecticides	Strontium
Chlorides	Iron, Lead	Sulfides
Chromium, Copper	Manganese	Sulfates
Cyanides	Nitrates	Sulfites
Dissolved gases	Nickel	Tars, Urea, Zinc
Detergents, Dyes	Organic chemicals	

A wide variety of microbes and microbial terminology are encountered in water pollution. The unicellular forms are quite important (Fig. 1). There are bacteria, fungi, protozoa, and algae. Forms of these are all active in stabilizing organic matter and even in removing toxic metals. Bacteria are usually less than 10 in size and are typically rod, spherical, or spiral shaped [Fig. 1(a)]. Many possess motile organs called flagella [Fig. 1(b)]. A typical microbial cell will contain a slime capsule, cell wall, cell membrane, a nucleus, and cell inclusions [Fig. 1(c)].

Like most bacteria, fungi are nonphotosynthetic and are typically much larger than the bacteria forming large mycelial strands [Fig. 1(d)]. They produce long stalks (sporophores) that bear and produce copious quantities of asexual spores. Many of the fungi reproduce by sexual as well as asexual means. Fungi tend to be filamentous and like bacteria use organic matter for sources of carbon and energy.

Algae are photosynthetic and they possess many diverse shapes and forms. Some of the free swimming ciliates are shown in Fig. 1(e).

The protozoa are also microscopic in size and feed heavily on bacterial populations, e.g., Sarcodina, Rotifers, Mastigophora, and some stalked ciliata.

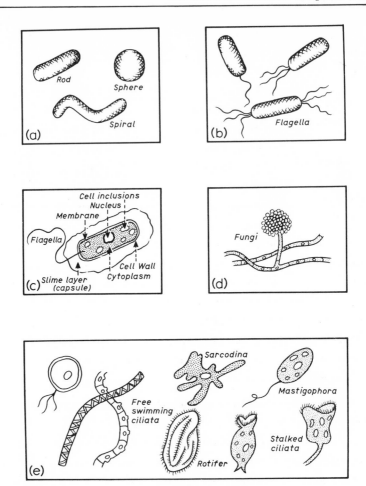

FIG. 1. (a) bacterial shapes; (b) motile organs on bacteria; (c) cell structure; (d) fungi; (e) algae (photosynthetic) and protozoa (nonphotosynthetic).

SUGGESTED REFERENCES

Handbooks

King, H. W. 1963. Handbook of Hydraulics. 5th ed., McGraw-Hill, New York.

Lange, N. A. 1961. Handbook of Chemistry. 10th ed., McGraw-Hill, New York.

Perry, J. H. (ed.) 1963. Chemical Engineer's Handbook. 4th ed., McGraw-Hill, New York.

Weast, R. C., S. M. Selby, and C. D. Hodgman, Handbook of Chemistry and Physics, The Chemical Rubber Co., Cleveland, Ohio.

Chemistry

Standard Methods for the Examination of Water and Wastewater. 1965. American Public Health Assn., New York, 12th ed.

Saywer, C. N. 1960. Chemistry for Sanitary Engineers. McGraw-Hill, New York.

Grit Chambers

Camp, T. R. 1942. "Grit chamber design." Sewage Works 14, 368.

Cawley, Wm. A. (Nov. 1966) "Sewage treatment plant equipment: grit removal." Water and Sewage Works, Ref. No. 113RN: R170-172.

Fair, G. M. and J. C. Geyer. 1954. Water Supply and Waste Water Disposal. Wiley, New York, pp. 610-615.

Sedimentation

Fair, G. M. and J. C. Geyer. 1954. Water Supply and Waste Water Disposal. Wiley, New York, pp. 601-615.

Sludge Digestion

Fair, G. M. and J. C. Geyer. 1954. Water Supply and Waste Water Disposal. Wiley, New York, pp. 755-792.

Sludge Digestion cont'd.

McCabe, J. and W. W. Eckenfelder, Jr. (eds.). 1958. Biological Treatment of Sewage and Industrial Wastes, Vol. 2, Anaerobic Digestion and Solids-
Liquid Separation. Reinhold Publishing Corp., New York.

Woods, C. 1966. "Evaluation of digester performance." Water and
Sewage Works Ref. No. 113 RN: R244-248.

Trickling Filters

Rich, L. G. 1963. Unit Processes of Sanitary Engineering. Wiley,
New York, pp. 47-56.

Rankin, R. S. 1956. "High-rate filters of biofiltration type and their
application to biologic treatment of sewage." In Biological Treatment of Sewage
and Industrial Wastes, Vol. 1. Ed. J. McCabe and W. W. Eckenfelder, Jr.
Reinhold, New York, pp. 304-324.

Smith, R. L. and T. Rodeberg. Nov. 1966. "A study of biological filter
media flora. Water and Sewage Works Ref. No. 113RN: R207.

Activated Sludge

Fair, G. M. and J. C. Geyer. 1954. Water Supply and Waste Water
Disposal. Wiley, New York, pp. 719-732 and 746-754.

McKinney, R. E. 1962. Microbiology for Sanitary Engineers. McGraw-
Hill, New York, pp. 213-237.

Haseltine, T. R. 1956. "A rational approach to the design of activated
sludge plants." Biological Treatment of Sewage and Industrial Waste. Vol.
1. Ed. J. McCabe and W. W. Eckenfelder, Jr. Reinhold, New York, pp.
257-271.

General

WPCF Manual of Practice No. 11. 1964. Operation of Wastewater
Treatment Plants. Water Pollution Control Federation, Washington, D. C.

General cont'd.

WPCF Manual of Practice No. 8. 1963. Sewage Treatment Plant Design. Water Pollution Control Federation, Washington, D. C.

Eckenfelder, Jr., W. W. 1966. Industrial Water Pollution Control. McGraw-Hill, New York.

Meers, J. E. and W. D. Straczek, 1967. "Basic calculations for waste-water treatment." Water and Sewage Works, 114-R. N., R254-R266.

Chapter 4

BENTHIC ORGANISMS AS POLLUTION INDICATORS

Biologists have discovered and classified a great wealth of information
concerning the anatomy, physiology, and morphology of countless marine
organisms of both the plant and animal kingdoms, However, the interests
of pollution engineers have been directed toward the mere presence of certain species and more recently to the relative numbers of individuals in each
species, particularly as indicators of aquatic pollution.

Although pollution is essentially a biological phenomenon in which the
biological flora and fauna in a system becomes upset, its primary effect is
most frequently assessed in chemical terms by measuring such characteristics as dissolved oxygen, pH, suspended solids, COD, ammonia, etc.
Relatively little concern is given to quantitating what has happened to the
biological system per se. Although chemical and physical methods have
traditionally been used in pollution assays because of their comparative ease
and directness, biological techniques have several practical advantages.
A single series of biological samples reveals the state of plant and animal
communities, which themselves represent a summation or static record of
prevailing and past environmental conditions since many pollutants are introduced intermittently into the water courses and may be missed by sampling programs other than by continuous automatic sampling.

Chemical and physical assays are prone to a wide range of results and
interpretations. In some instances an average result may not be truly indicative of the situation, whereas biological studies show up the cumulative
effects of such intermittent discharges or the effects of a single, highly
toxic discharge, which chemical analysis may miss. Conversely, biological
results are not usually markedly affected by a temporary amelioration, nor
by a transient deterioration of the effluent.

The wide range of characteristics that must be evaluated in a physical
and chemical program necessitates a large expenditure for laboratory

73

equipment; however, a bioassay can usually provide an equally good or better result for a fraction of the cost.

A chemical assay can detect the presence of certain pollutants, but it is possible to determine the source only if the discharge is continuous. The biologist, on the other hand, needs only to move upstream to where the effects begin. Nowak (1940), Thomas (1944), and Huet (1954) demonstrated that chemical data gave a good indication of the biological state of water only if interpreted with due care and considered with a number of chemical and physical factors. Butcher (1946) showed that chemical and biological data may differ widely, particularly when several types of pollutants are involved, and asserted that pollution should be defined in terms of biological conditions rather than by chemical standards.

Beak (1964) indicates that the biological scores from two tests in October of 1958 and 1959 compared fairly well with the results of 17 chemical surveys done in each of those two years. The results were such that an extensive and expensive series of chemical tests could be replaced by a single biosurvey with a concomitant saving in costs and no decrease in the quality of the results of the assay.

The biological effects of pollution are usually twofold: (a) the elimination of certain intolerant species due to the resulting decrease in competition and a proliferation of the species remaining, and (b) the replacement of normal, clean-water communities by others more adapted to the particular ecological conditions induced by a pollutant(s).

An examination of the literature indicates that there is dispute among the biologists as to which type of organism is the best indication of pollution. Much work (Redeke, 1927; Ellis, 1937; Cole, 1941; Doudoroff and Katz, 1950) has been done to determine the relative sensitivity of various fishes to pollutants and to ascertain the tolerable levels of toxicity. Even more important is the difficulty in catching and sampling fishes in statistically significant numbers.

Liebmann (1942) claims that the microscopic benthic organisms are the most useful as indicators of pollution on grounds that they are more generally distributed and that smaller organisms react more readily to changes in the environment because the surface area to mass ratio increases with diminishing organism size (Fig. 1). This author also pointed out that in zones of severest pollution microscopic organisms are present where even the most durable of the macroinvertebrates are scarce or absent. The suitability of microbiota is diminished unfortunately because of their rapid rate of reproduction and the difficulty of correctly identifying species. Such an ability generally comes only with years of training and experience. Furthermore, it is difficult to sample microorganisms, many of which must be examined alive for purposes of proper identification. With protozoa, identification must be accurate in that very similar genera occur under quite different saprobic conditions (Liebmann, 1951). Not enough is known, however, about the reaction of nonpathogenic microorganisms in polluted waters or how to assess their meaning. Information will gradually be developed in this area, but it will be slow because microbial taxonomic studies are time consuming and expensive.

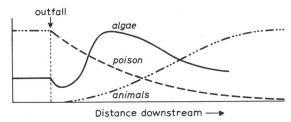

FIG. 1. Effect of pollutant on unicellular algae or microbes and on benthic organisms.

As a result of their relative lack of mobility, their less rapid rate of reproduction, the ease of collecting statistically significant numbers, and the relative ease of identifying them, particularly after preservation, the benthic macroinvertebrates are generally regarded as the most suitable indicators of pollution. Several investigators (Richardson, 1921 and 1929; Gaufin and Tarzwell, 1952, 1956; Huet, 1949; Brinkhurst, 1966; and Wilhm and Dorris, 1966) have relied almost entirely on them. Certain species of sponges, for example, respond to various type of poisonous pollutants in even very mild cases, while others (sludgeworms and maggots being the best examples) can tolerate even the most gross organic pollution and high levels of toxic pollution. Worms exhibit very marked physiological tolerance for oxygen depletion related to excess decomposible organic matter present in the environment, but they do decrease in number when conditions are at their worst. Few other organisms can survive under these circumstances, so that worms, which have a very efficient oxygen uptake mechanism, may make up the entire benthic community (Fig. 1).

Measurements that summarize community structure clearly and briefly are valuable in evaluating the effects of organic enrichment. To this end various attempts have been made to quantify, correlate, and systematize data obtained from river bottoms in an effort to obtain clear cut indication of pollution from the number of individuals and species present per unit area in a given situation. Fisher et al. (1943) described the benthic community structure in terms of a logarithmic series. Preston (1948) suggested a log-normal distribution. Here, the number of individuals in each species is plotted on a logarithmic scale against the number of species on a linear base to give a "normal" distribution curve. Gleason (1922) proposed a linear relationship between the number of species and the logarithm of the area, whereas Margalef (1956) indicated that the number of individuals, under particular conditions, was directly proportional to the area.

Margalef (1956) suggested that previous mixed-species analyses were dependent upon sample size and proposed that the relationship between the total number of species and the total number of individuals has a greater significance. It was indicated that under clean-water conditions many species were present, whereas in polluted waters fewer species with relatively more individuals per species could be identified. With many species present and relatively few individuals per species, the sample is described using information theory, which has been developed elsewhere herein as having a high diversity, while a community with many individuals representing a few species is described as having a high redundancy.

Attempts have been made (Shrivastava, 1962) to define environmental requirements for certain species, but most species are quite cosmopolitan and are to be found in a large variety of habitats differing in current velocities, sediments, bottom materials, and other physicochemical factors. Another approach (Brinkhurst and Kennedy, 1965) is a chemical and micro-biological examination of food sources and their uptake by the worms, as it was thought possible that each species utilizes a specific type of organic trophic constitutent of the benthos either directly or via its microflora and fauna. Preliminary studies indicate that even in the sources investigated

there appears to be no critical or significant differences in the food requirements of separate species. The trophic chains of organisms involved with
changes in substrates are probably overlapping and difficult to detect.

Different benthic species of course have different ranges of tolerance to
organic wastes (Richardson, 1928) and several investigators (Gaufin and
Tarzwell, 1952; Surber, 1953; and Mackenthum et al., 1956) have tried to
classify such organisms according to pollution tolerance. The "indicator"
species approach must be used with caution because many of these species
occur under both polluted and unpolluted conditions.

Another system for biological assessment of pollution was proposed by
Patrick (1950, 1951) and modified by Wurtz (1955). This involves examination of a number of stations on healthy reaches of a river and counting of the
number of species present in each of seven groups. The assumption here is
that each group is a unit in which those species that are grouped together
behave in the same definitely recognizable way under the influence of a pollutant. Using this method, tests are also made in the polluted reaches of the
river, and the number of individuals in each species is compared statistically with the standard from a healthy portion and expressed as a percentage. This approach is very cumbersome and time consuming and requires
accurate identification of each species.

Various investigators have identified or delineated several zones in
rivers based on the severity of pollution each with its characteristic flora
and fauna. Kolkwitz and Marsson (1908, 1909) defined (a) a polysaprobic
zone characterized by gross pollution with organic matter of high molecular
weight, very little dissolved oxygen, the presence of sulfides, abundant
bacteria, and the presence of a few invertebrate species living on decaying
organics and bacteria; (b) a mesasaprobic zone characterized by the presence of simpler organic molecules, increasing dissolved oxygen, many
bacteria, fungi, few algae, and several species of the more tolerant animals,
and (c) an oligosaprobic zone where mineralization is complete, oxygen
levels are normal, and a wide range of plants and animals is present. A
large number of the flora and fauna that were tested were characteristic of
each zone, and identification of those present in a river made it possible to
establish its pollutional status. Liebmann (1951) devised a modification of
this system predicted on the same basic idea, but with a more careful

selection and more detailed description of organisms, including notes on abundance and the particular conditions under which various organisms thrive.

The most recent and currently acceptable approach for evaluation of the degree of pollution has been advanced by Wilhm and Dorris (1966), who proposed the so-called information theory approach. The designation of particular species as "indicators" of pollution requires specific identification of orgamisms and of course requires special taxonomic skills. Information theory enables investigators of limited skill to evaluate stream conditions since less precise taxonomic distinctions need be made. The only data required for community structure analyses are the total number of taxa present in a given area and the number of individuals in each. The assignment of scientific names is not necessary.

A system of quantifying biological data, as used by Beak (1964), entails dividing species into three groups: (a) those that are quite tolerant of polluted conditions and often occur in large number under such circumstances, (b) those that are moderately tolerant of pollution and exist in smaller numbers in both mildly polluted and unpolluted areas, and (c) those species that are intolerant of pollution and do not occur in polluted locations. The species used by Beak are quite common and readily identifiable. A biological score is derived for each investigation site based on the presence or absence of species. These are usually correlated with a series of tests done for dissolved oxygen, BOD, and suspended solids.

Pollution of water courses can generally be classed in one of six categories: (a) poisons, (b) suspended solids, (c) organic matter, (d) nonpoisonous salts, (e) simple deoxygenators, and (f) heat. Each type of pollutant has a characteristic effect on the flora and fauna of the watercourse.

Perhaps the most serious effects on the normal ecology of rivers is brought about by toxic industrial wastes and mine effluents containing acids, alkalies, heavy metal ions, cyanides, phenols, insecticides, etc. An undue abundance of worms in relation to the arthropods in a community together with a wide diversity of species of worms is often indicative of pollution by insecticides that may enter the watercourse through indiscriminate aerial spraying. The effects of such poisons become progressively less severe as one proceeds downstream from the point of origin because of neutralization precipitation, adsorption, oxidation, or dilution. These toxicity gradients have a corresponding inverse gradient on the number of species present. Near the outfall of an effluent the number of both plant and animal species is reduced in proportion to the severity of toxicity.

Jones (1940) found that in parts of the Rheidol River all species of worms, leeches, crustaceans, molluscs, and fishes had been eliminated by drainings from lead and zinc mine dumps. Pentelow and Butcher (1938) found a similar elimination of all fauna due to copper pollution. The biological effects of pollution by toxic substances are essentially the simple elimination of certain species for varying distances below the outfall. If conditions are not unsuitable surviving species may become unusually abundant because of the absence of competition or other biotic factors that normally control them. There appears to be no specific biotic community associated exclusively with any particular type of toxic substances.

Concentrations of suspended solids decline downstream from the outfall due to differential settling rates. The settling rate in a given fluid depends on the density, shape, and size of the particles and the velocity of the flow. Fine solids remaining in suspension have very little direct effect on the fauna, but they do screen out sunlight, resulting in reduced growth of plants on which benthic organisms feed. Surber (1953) reported the elimination of all the flora and all but two species of fauna (which fed on detritus and vegetable matter of terrestial origin) in the Menominee River, which had been contaminated with fines from an ocher mine. Where settling occurs, the sediments fill the interstices among the stones, thereby depriving cryptic organisms of hiding places. Clean water invertebrates decline rapidly because of the lack of oxygen and the blanketing effects of suspended solids and sewage fungi. Stone surfaces are covered, rendering ineffective the various hold-fast mechanisms of stone-fauna. Consequently, most of the normal or typical fauna disappear, to be replaced by burrowing and tube-building species such as the Oligochetae worms and Chironomid larvae.

Organic residues in suspension or in solution are produced by most of the food processing industries, laundries, and flour, paper, and textile mills. Domestic sewage is a very important source of such effluents, the chief characteristic of which is that they are chemically unstable, readily oxidizable and, in fact, remove large amounts of dissolved oxygen from the water (O_2 sag). The extent of the deoxygenation depends on the nature and amounts of the pollutant, the volume and velocity of flow of the stream, and the initial oxygen content. In extreme cases a septic condition may result, with the complete depletion of oxygen and the production of methane and sulfides. As the organics are being decomposed, ammonia and phosphates (Fig. 2) are released, producing a series of complex changes succeeding one another downstream.

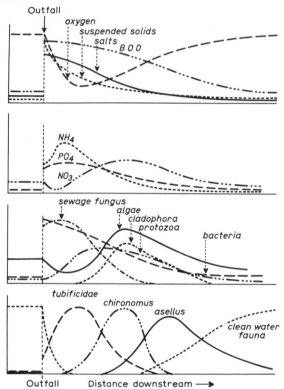

FIG. 2. General influence of organic pollutants on flora and fauna (Hynes, 1959).

The biological effects are correspondingly complex with those organisms, such as the gill breathing nymphs of stoneflies and mayflies, which are inefficient at extracting dissolved oxygen from the water and are sensitive to pollutants and disappear first. Bacteria decline in number as the organic matter is utilized. Protozoa appear, growing on bacteria and then decline. Algae at first decrease but, as nutrient salts are released from the degrading organics and conditions improve, they increase greatly in abundance. Among the invertebrates (Table 1) are found the larvae of Eristalis, Psychoda, and Culicine mosquitos in grossly polluted zones and the Tubifex and Limnodrilus hoffmeisteri worms, with the leech Trocheta subviridus and the larvae of Chironomus thummi found further downstream.

TABLE 1

PER CENT COMPOSITION OF FAUNAS OF TWO ERODING BRITISH RIVERS, THE GREAT DRIFFIELD, A "HARD" RIVER, AND THE AFON HIRNANT, PRESENTED HERE TO INDICATE THE FAUNA COMMONLY PRESENT AND USED IN BENTHIC STUDIES OF RIVERS[a]

Common class	Genera or Family	Great Driffield	Afon Hirnant
Flatworms	Tricladida	1%	2.5%
Worms	Oligochetae	27	2.5
Leeches	Hirudinea	1	--
Crustacea	Shrimps (Amphipoda)	32	--
	Asellus	2.5	--
Mayflies	Baetis	3	5
	Caenis	--	--
	Ephemerella	2.5	4
	Ephemera	--	--
	Ecdyonuridae	--	19
Stoneflies	Leuctridae and Vemouridae		47
	Perlidae and Perlodidae	--	2
	Chloroperla	--	2
Caddis-worms	Caseless	--	9
	With cases	30	2.5
Bugs	Hemiptra	--	--
Beetles	Helmidae	1	1
	Other beetles	1	1
Midges	Chironomidae	2	4
	Ceratopogonidae	2.5	--
Buffalo-gnats	Simulium	2.5	6
Water-mites	Hydracarina	1	2.5
Snails	Gastropoda	2.5	--
Limpets	Ancylastrum	2.5	2.5
Other flies	Diptera	1	2.5
Other animals, mostly insects		1	2.5

[a] After H.B.N. Hynes, 1966.

All these organisms are not normal clean-water denizens, although they may appear in great numbers. Richardson (1929) reported 3.5×10^5 Tubifex worms per square yard, while Meschkat (1937) reported a density of $1.7 \times 10^6/m^2$ of the same species in polluted reaches of the river Elbe. Gradually, as the organic matter is decomposed and the stream is reoxygenated, the normal clear-water fauna reappear.

Effluents containing nonpoisonous salts are typical of mining and oil-drilling operations. Lafleur (1954) found that fishes are generally more tolerant of salts than are the invertebrates, but certain species, among them crayfish and dragonfly nymphs, are able to withstand high concentrations of salts, while others, including several diatoms, the brine shrimp Artemia salina, and the salt-fly Ephydra riviparia, thrive in saline habitats. The presence in inland waters of certain species normally found in salt or brackish localities where salinity fluctuates or is high may indicate enrichment of human activity. Few species can adapt to saline conditions and saline pollution therefore may result not only in the elimination of certain species but their replacement by others. Classen (1926), in a study of what had originally been a typical trout stream but was polluted with sodium chloride, observed fauna including the saltfly Ephydra, biting mosquitos, Eristalis, a few worms, and beetle larvae. Other common river fauna are shown in Table 1 (Hynes, 1966).

Heat, a progressively more common and more significant form of pollution, usually enhances the effects of chemical pollutants (Fig. 3). It increases the toxicity of poisons. Herbert et al. (1955) have shown that a rise of $10^{\circ}C$ halves the survival time of test animals. It increases the metabolic rate of microorganisms and it stimulates all biological activity, thereby increasing the oxygen demand and the oxygen sag. This causes a marked deterioration of the environment. This is associated with increased metabolic activities and a shortening of the length of the successive biological zones because the organic matter is more rapidly destroyed. Tuxen (1944) has shown that in springs hotter than $40^{\circ}C$ certain groups — namely, Protozoa, Rotifera, Nematoda, Annelida, and Ostracoda -- are frequently present, while others, including Tiprilidae, Simuluda, Culicidae, Psychodidae, and Trichoptera, are always absent. By analogy, a similar limited selection of organisms might be expected to occur in thermally polluted waters.

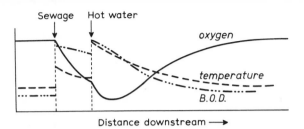

FIG. 3. The effect of heated effluents on organically polluted rivers.
(After Hynes, 1959.)

From the foregoing it can safely be concluded that relationship of the
benthic organisms with their environment is very complex and is influenced
by many factors, all of which must be considered when making evaluations of
pollution in terms of the flora and fauna content. Factors such as the type of
river bed and the flow velocity must be considered. Some species prefer
slow waters to avoid being scoured and swept away. Some organisms are
bottom dwellers and feed off the surface, while others are burrowing by pref-
erence. Thus, whether a river is eroding or depositing will significantly
alter the dominant species present and their relative distribution. The hardness of the water also affects the kinds and numbers of species present. As
water becomes harder or more alkaline, a dominant biological population
becomes established.

Despite the multiplicity of interacting factors, a study of the benthic
organisms of an area in terms of the presence or absence of certain species
or of their associations can give an accurate estimation of the degree of pollution present. Surveys should be made of waters before engineering or disposal works are constructed to determine the normal clean-water denizens
of an area, and this should be repeated after construction and operation to
assess the extent of pollution or environmental alteration. Such surveys are
necessary and should be compulsory. In such analyses the effects on the
benthic macroinvertebrates are considered to be the most significant.

REFERENCES

Beak, T.W. 1964. "Biological measurements of water pollution." Chem. Eng. Papers, 60, 33-45.

Brinkhurst, R.O. and C.R. Kennedy, 1965. "Studies on the biology of the Tubificidae in a polluted stream." J. Animal Ecology, 34, 429-443.

Brinkhurst, R.O. 1966. "Detection and assessment of water pollution using oligocheate worms (Parts 1 and 2)." Water and Sewage Works, Oct. & Nov. 113, 398-401 and 438-442.

Butcher, R.W. 1946. "The biological detection of pollution." J. Inst. Sewage Purif. 2, 92-97.

Classen, P.W. 1926. "Biological studies of polluted areas in the Genesee River system." Ann. Rep. N.Y. Conserv. Dept. 16, Suppl. 38-47.

Cole, A.E. 1941. The Effects of Pollutional Wastes on Fish Life: A symposium on Hydrobiology. Madison, London. 241-259.

Doudoroff, P. and M. Katz, 1950. "Review of literature on the toxicity of industrial wastes to fish." Sewage & Industrial Wastes, 22, 1432-1458.

Ellis, M.M. 1937. "Detection and measurement of stream pollution." Bull, U.S. Bureau of Fisheries, 48, 365-437.

Farrell, M.A. 1931. "Studies of the bottom fauna in polluted areas." N.Y. Conserv. Dept. Biol. Survey, No. 5 Suppl. 20 Annual Rept. 192-196.

Fisher, R.A., A.S. Corbet, and C.B. Williams, 1943. "The relation between the number of species and the number of individuals in a random sample of the animal population." J. Animal Ecology, 12, 42-58.

Gaufin, A.R. and C.M. Tarzwell, 1952. "Aquatic invertebrates as indicators of stream pollution." Public Health Report, 67, 1, 57-67.

Gaufin, A.R. and C.M. Tarzwell, 1956. "Aquatic microinvertebrate communities as indicators of organic pollution in Lyttle Creek." Sewage & Ind. Wastes, 28, 906-924.

Gleason, H.A. 1922. "On the relation between species and area." Ecology, 3, 158-162.

Herbert, D.W.M., K.M. Downing, and J.C. Merkens, 1955. "Studies on the survival of fish in poisonous solutions." Verk. int. Ver. Limnol. 12, 789-794.

Huet, M. 1949. "La pollution des eaux. L'Analyse biologique des eaux polluees." Bull. Centr. Belg. Etude Doc. Eaux. 5 and 6, 1-31.

Huet, M. 1954. "Biologie, profils en long et en travers des eaux courantes." Bull. franc. Piscic. 175, 41-53.

Hynes, H.B.N. 1966. The Biology of Polluted Waters. Liverpool University Press, Liverpool.

Jones, J.R.E. 1940. "The fauna of the river Melinder, a lead-polluted tributary of the river Rheidol in north Cardigonshire, Wales." J. Animal Ecol. 9, 188-201.

Kolkwitz, R. and M. Marsson, 1908. "Okologie der pflanzlichen Saprobien." Ber. deutsch. bot. Ges. 26, 505-519.

Kolkwitz, R. and M. Marsson, 1909. "Okologie der tierische Saprobien." Int. Rev. Hydrobiol., 2, 126-152.

Lafleur, R.A. 1954. "Biological indicies of pollution as observed in Louisiana streams." Bull. La. Eng. Exp. Station, 43, 1-7.

Liebmenn, H. 1942. "Die Bedeutung der mikroskopischen Untersuchung fur die biologische Wasseranalyse." Vom Wasser, 15, 181-188.

Liebmann, H. 1951. Handbuch der Frishwasser und Abwasserbiologie. Munich, Oldenbourg.

Lowndes, A.G. 1952. "Hydrogen ion concentration and the distribution of freshwater entomostraca." Ann. Mag. Nat. Hist. 12, 58-65.

Mackenthum, E. W., W. M. Van Horne, and R. F. Balch, 1956. "Biological studies of the Fox River and Green Bay." Report submitted to the Division of Water Pollution Control, Wisconsin.

Margalef, R., 1951. "Diversidad de especies en las comminudedes naturaks." Pub. Inst. Biof. Apl. (Barcelona) 9, 5-27.

Margalef, R. 1956. "Informacion y diversidad especifica en las communidades de organismos." Invest. Pesq. 3, 99-106.

Meschkat, H.H., 1937. "Occurrence of an exotic oligochaete Bronchiura sowerbyi in the River Thames." Nature, London, 182, 732-38.

Nowak, W., 1940. "Uber der Verunreinigung eines kleinen Elusses in Mahren durch Abwasser von Wussgerberein, Leder-Leimfabriken und anderen Betrieben." Arch. Hydrobial (Plankt), 36, 386-423.

Patrick, R., 1950. "Biological measure of stream conditions." Sewage & Ind. Wastes, 26, 926-938.

Patrick, R. 1951. "A proposed biological measure of stream conditions." Verk. int. Ver Limnol, 11, 289-307.

Pentlow, F.T.K. and R.W. Butcher, 1938. "Observations on the condition of the Rivers Churnet and Dove in 1938." Report Trent Fisheries Dist. App. 1.

Pentelow, F.T.K. and R.W. Butcher, 1938. "Observations on the condition of rivers Churnet and Dove in 1938." Ann. Rep. Trent Fish. Bd. Nottinghm, App. 1.

Preston, F.W. 1948. "The commonness and rarity of species." Ecology, 29, 254-283.

Redeke, H.C. 1927. "Report on the pollution of rivers and its relation to fisheries." Rapp. Cons. Explor. Mer. 43, 1-50.

Richardson, R.E., 1921. "Changes in the bottom and shore fauna of the middle Illinois River as a result of increase southward of sewage pollution." Bull. Ill. Nat. Hist. Surv. 14, 33-75.

Richardson, R.E. 1929. "The bottom fauna of the middle Illinois River." Bull. Ill. Nat. Hist. Surv. 17, 387-475.

Shrivastava, H.N. 1962. "Oligochaetes as indicators of pollution." Water and Sewage Works, 109, 40-41 and 387-389.

Surber, E.W. 1953. "Biological effects of pollution in Michigan waters." Sewage & Ind. Wastes, 25, 79-86.

Thomas, E.A. 1944. "Versuche uber diefelbstreinigung fliessenden Wassers." Mitt. Lebensm. Hyg. Bern. 35, 199-216.

Tuxen, S.L. 1944. "The hot springs, their animal communities and their zoogeographical significance." The Zoology of Iceland, 1, 2, Munksgaard.

Wilhm, J.L. and T.C. Dorris, 1966. "Species diversity of benthic macroinvertebrates in a stream receiving domestic and oil refinery effluents." Am. Midland Naturalist. 76, 427-449.

Wurtz, C.B. 1955. "Stream biota and stream pollution." Sewage & Ind. Wastes. 27, 1270-78.

Chapter 5

STREAM SURVEILLANCE

No one has described an ideal stream surveillance program. Definition of such programs are needed because without quantitative chemical, physical, and biological data stream norms cannot be established. Changes in our streams are occurring so rapidly because of increased urbanization, industrialization, and new farming methods that data are needed for establishing controls. This information will also teach us what effects chemical and physical changes have on the biological environment.

Objectives of stream surveillance programs are (a) to determine the effect of domestic and industrial waste discharged, (b) to assess the suitability of water for domestic and industrial supply, (c) to assess the environmental characteristics in terms of protecting the fish for recreational and commercial purposes, (d) to determine the hydrological characteristics in terms of dam building and creation of reservoirs, and (e) to assess the contribution of various waste sources in terms of the contribution the stream makes to the receiving body.

I. STREAM CLASSIFICATION

The necessity for protecting stream uses to give maximum benefit can be resolved by a program of classification whereby water uses are defined for each stream. The resulting pollution abatement program is a direct reflection of this classification. Dappert (1958) states that the New York State Water Pollution Control Board gives consideration to the following characteristics in its classification of streams:

1. The size, depth, surface area coverage, volume, direction and rate of flow, stream gradient, and temperature of the water.

2. The character of the district bordering said waters and its peculiar suitability for the particular uses of residential, agricultural, industrial, or recreational purposes.

3. The uses which have been made, are being made or may be made of said waters for transportation, domestic and industrial consumption, bathing, fishing and fish culture, fire prevention, the disposal of sewage, industrial wastes, and other wastes.

4. The extent of present defilement or fouling of said waters which has already occurred or resulted from past discharges therein.

The time factor involved in obtaining this information is considerable, and therefore surveillance programs are an absolute necessity. Water uses and requirements change very quickly in an increasingly urbanized and industrialized society and the danger of initial classification with insufficient data and knowledge of future requirements is very real. The collection of an inventory of background data is an essential part of stream management. All water resources commissions are presently engaged in programs of data collection. Emphasis is not always placed on classification of stream resources but rather on the development of realistic objectives based on present data and the systematic revision of these objectives in the light of future data and changes in water use.

II. PHYSICAL CHARACTERISTICS OF STREAMS

The variation of flow with changing seasons is one of the most complicated problems of river hydrology. Flow variation in a river is superficially
simple but difficult to describe quantitatively. Seasonal variations and why
they occur can also be very difficult to describe quantitatively. They are a
combination of the effects of topographical relief and climate (Browzin, 1962).

River Hydrographs

River flow is variable in space, changing from the source to the mouth.
The problem is to find a common characteristic that can be used to reflect
these variations in time and space. The "river hydrograph" is used to resolve this problem.

River hydrographs are computed monthly from flow data averaged over
a period of many years. For example, if the average monthly flow of a river
for May of a particular year is 1780 ft^3/sec and the average mean flow of the
river is equal to 525 ft^3/sec, the coefficient of flow is $\frac{1780}{525}$ = 3.40. This
coefficient reflects the flow conditions during the month of May. (This concept of monthly flow coefficients was first proposed by the French hydrologist
Coutagne.)

Browzin (1962) classified rivers in the Great Lakes - St. Lawrence Basin
according to monthly flow coefficients. Along the south shore of Lake Erie
(Figure 1) the streams have very low flow coefficients (0.05-0.10) from August to October. High flow coefficients occur from January to March. Peak
flow never occurs in February. Unlike streams along the south shore of Lake
Erie, the streams on the north shore and extending along the Niagra Peninsula
do not show the extremely low flow coefficients. Values do not drop below
0.30 and high flows occur later in March and April. The pattern changes again
for streams east of Lake Erie which run into Georgian Bay (Fig. 1). In
spite of the presence of many lakes in the stream systems which might be expected to exert a moderating effect on stream flows, the streams are apparently not regulated by them. The flow regime is more severe than it is around
the Niagara Peninsula, with maximum stream coefficients going as high as
3.40 and occurring one month later due to the more northerly location. Spring
floods result from melting snow and a second rise in flow rate occurring in
late fall is due to the increase in evapotranspiration and to an increase in
rainfall in winter in certain basins.

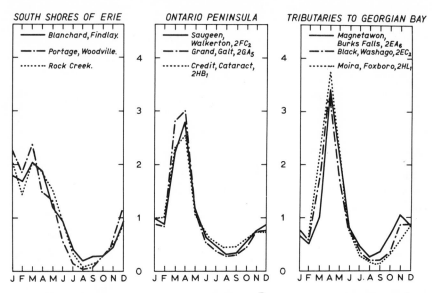

FIG. 1. River hydrographs of streams. [From Browzin (1962), "On classification of Rivers in the Great Lakes - St. Lawrence Basin.]

III. SAMPLING PROGRAMS

The collection of reliable water quality data begins with sampling. The sample by definition should be representative, for it is presented as evidence of the quality of water from which it is obtained. The variations encountered in chemical and biological assay programs and how to establish sample stations and take samples are difficult problems (OWRC, 1967; Hoskins, 1938; Hynes, 1963; Johnson, 1964, 1966; Thomas, 1953; Welch, 1948).

The location of sampling stations in the design of a stream surveillance program is largely one of common sense and experience in predicting where the major changes in the stream are taking place. Stations should be located above and below major waste inputs in the stream. Measurements should also be made at outfall pipes at specific distances from the pipe. The pattern of flow from an outfall should be determined by the injection of rhodamine or other dyes into the outfall pipe. Major attention should also be given to stations below the inflow of tributaries into a stream. The temptation to locate stations at bridges and areas with easy access to the stream is great. Since access to streams with the necessary sampling and equipment for analysis is a legitimate problem, a compromise between the access factor and significance can usually be reached.

One problem is establishing the location of outfalls. The rhodamine dye technique mentioned earlier for establishing the dispersion pattern of effluents is also an excellent technique for locating the pipe. Where the use of dye raises objections to the impairment of the aesthetic quality, the use of a temperature probe operated from a boat will locate the source of the warmer effluents quickly.

Once a preliminary station location grid has been sampled a number of times, statistical tests should be done on various parameters to establish whether the changes between the stations are significant. If they prove to be insignificant, the station should be deleted. The cost of a sampling station is high and should not be maintained unless needed. A concentration of activity on a limited number of stations where significant changes are taking place is preferable to spending the same time on more stations. These so called "monitoring" stations are backed up by at least one detailed survey per season including many stations to see whether additional changes in the stream are occurring.

Considerable controversy exists between engineers and biologists on the respective merits of chemical and biological sampling. Most of the conclusions on this subject are that the two techniques compliment each other.

IV. CHEMICAL PROGRAMS

A. Advantages

Chemical studies far outnumber biological studies for two reasons: (a) chemical data are readily quantified and are easily analyzed statistically; (b) computer analysis is easy.

Chemical techniques give an indication of the water quality at one particular point at a particular moment in time. Use of a series of sampling points down a stream can provide information on many relationships.

B. Oxygen Relationships

Dissolved oxygen content in streams can be described as the resultant of two forces. The first seeks to deoxygenate a stream in order to satisfy its oxygen demand, and the opposing force is active in reoxygenation to the level of DO saturation at the prevailing temperature of the stream and the partial gas pressure of the overlying atmosphere. Organic pollutants or reduced inorganic effluents tend to lower the DO content of a stream, producing what is known as an oxygen sag curve (Dawson, 1962) (See Fig. 2).

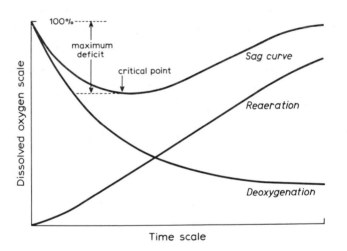

FIG. 2 Deoxygenation, reaeration, and sag curve from Dawson (1962).

The maximum deficit and the critical point or low point in the oxygen sag curve give an indication of the oxygen requirements of a stream.

C. Dilution

Chemical measurements on a series of stations also give an indication of the dilution capacity of a stream when conservative parameters are used.

These parameters are used for chloride, iron, copper, nickel, and zinc contents of a stream.

D. Nutrient Content

Chemical measurements also provide valuable information on the nutrient content of the stream. The nutrients most commonly measured are nitrogen and phosphorus. Measurements on all compounds in the nitrogen cycle - namely, Kjeldahl or organic nitrogen, nitrites, and nitrates - give an indication of the degree of nitrification that is taking place.

V. BIOLOGICAL PROGRAMS

One might ask "Why involve biologists in pollution studies when physical and chemical methods provide a quick easy way of quantifying stream conditions?" The reasons are quite practical:

1. Chemical and physical measurements are prone to wide fluctuations, an average of which provides a biased view on the actual state of pollution.

2. Many pollutants are released intermittently and may be missed by all sampling programs with the exception of continuous automatic monitoring.

3. The range of substances that must be tested in a chemical program requires a huge outlay of equipment in the laboratory. Biological techniques can often provide an equally good answer at lower cost.

4. The presence of some chemically identified pollutant often does not betray the source of pollution; it just indicates that the flow of the stream carried it to the sampling location.

5. Helps relate pollutant to its toxic properties.

Biological techniques of assessing pollution have not been accepted readily by the scientific community for the following reasons:

1. Biologists have been reluctant to present their results in a form that is readily comprehensible to other scientific disciplines.

2. Biologists have been unable to come to agreement among themselves on standard techniques for the assessment of pollution.

3. Biologists have failed to emphasize the importance of systematics in water pollution work, and when this is known, they have failed to make it readily available.

A. Biological Techniques

The main advantage of the biological techniques is that, unlike chemical techniques, it measures changes in the environment. One biological survey can therefore show the results of the variations in chemical and physical conditions that have occurred in the survey area for a period previous to sampling.

The basis of the technique is that in a normal unpolluted river there will be a community of animals and plants in a state of equilibrium with the environment. The number of species and number of individuals in each species will be dependant on a number of environmental factors, mostly concerned with the chemical and physical characteristics of the water and the nature of the bottom deposits. Usually this environment will include a number of species that are tolerant of environmental change and others that are very intolerant of such change. See Chapter 4 for discussion of biological indicators of pollutions and bioanalytical approaches to pollution.

Brinkhurst (1966) states that the emphasis of biological pollution studies must be on communities of animals. Emphasis was at one time placed on the presence or absence of an individual species called a "biological indicator" that would provide an indication of pollution. This concept is based on the premise that there are species of animals or plants that inhabit only polluted conditions. The search for indicator species was prompted by the desire to find a simple method of biological measurement of pollution requiring a minimum of biological knowledge. This approach has not met with success.

Biological studies of polluted environments identifying only animals to the level of the family or cited proportions of certain worms to insects or other invertebrate forms are quite insubstantial. Such crude methods will only reveal gross alterations of a habitat, of the type that are easily established by simple visual inspection.

B. Choice of Biological Investigation

The question arises as to which group of organism one should us to assess the degree of pollution. Ideally, one should use fringe, bacteria, algae, plants, and animals, but biologists necessarily tend to use specialized groups, whether they be fish, bottom organism, or algae.

Fish are difficult to use because they are difficult to sample and their mobility tends to simply make them avoid polluted areas in streams. Changes

in fish speciation in the streams are common. These changes are often due
to the killing of fishes by subtle environmental changes affecting eggs and
early developmental stages of fish species.

Doudoroff (1951) states that biological studies of polluted environments
should be centered on the fishes and organisms that fish consume. Biological
studies of industrial pollution of water should be directed primarily toward
reliable estimation and understanding of the effects of pollutants on aquatic
resources of economic and recreational input. Industrial waste disposal
should be controlled so as to avoid any direct or indirect damage to these
resources.

These objectives often cannot be achieved through studies of those aquatic
organisms that are of no immediate value to man and that are considered only
as indicators of stream conditions. When it is not necessary, inferential
estimation of pollutional damage to fish life is discouraged and a direct approach to the evaluation of fish habitats and their productivity, and of the
toxicity of wastes is recommended. Comparative studies of fish population
and fishing success in comparable polluted and unpolluted waters are most
pertinent. Studies of the bottom fauna (benthos) and of plankton also can be
very instructive, but the relative importance of these various organisms as
food for desirable fish species should be a major consideration in the conduct
of such studies.

Diatoms have been extensively used by certain workers especially Patrick
(1953, 1956). They are suitable because:

1. They need no special treatment for preservation because the cell wall on
which the identification is based is composed of silica.

2. The diatom flora of a normal stream is made up of a great many species
and a great many specimens. Thus the group lends itself to statistical analysis.

3. Diatoms have a great range in sensitivity to chemical and physical conditions of water. The species are abundant enough so that they can be used
as an indication of most types of aquatic environments.

4. A great deal of information is already available as to the type of environment in which many species are found.

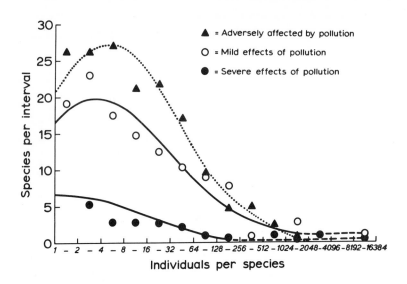

FIG. 3. Graphs of diatom populations in rivers showing varying degrees of pollution. From Patrick (1966), "Diatoms as indicators of changes in environmental conditions."

A comparison (Fig. 3) showing the character of diatom populations from unpolluted, mildly polluted, and extremely polluted situations is shown by Patrick (1956). A progressive decrease in the number of species and an increase in the number of individuals in each species is evident with increasing pollution. The diatoms and other types of microflora unfortunately require specialists to identify them, and the results of a study may in certain cases reflect only local conditions because of the rapidity in which these communities may develop.

The usefulness of macrobenthos communities involving worms, snails, clams, insects, etc. in providing an indication of the general condition of environments has been reported by Beak (1958, 1964) and Brinkhurst (1966) in Canada (see Chapter 4). These animals may better reflect the condition of the environment due to (a) their relative lack of mobility (in contrast to fish), and (b) their less rapid rate of reproduction (in contrast to the microbiota).

Oligocheate worms, and in particular the Genus Tubifex commonly known as sludge worms, are used. Beeton (1965) has done a study of the bottom

fauna of western Lake Erie utilizing samples gathered in 1930 and 1961 to show
the progressive decrease of Hexagenia nymphs (more commonly known as
mayfly nymphs), a group very nonresistant to organic pollution and the progressive increase of oligocheate or sludge worms (Fig. 4).

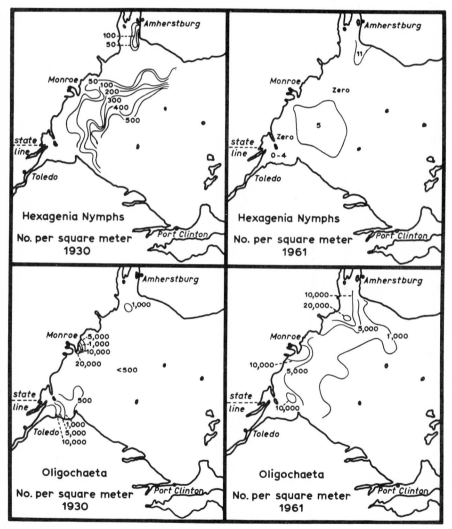

FIG. 4. Distribution and abundance of Hexagenia nymphs and oligochaetes
in Western Lake Erie, 1930 and 1961.

C. Quantifying Biological Data

Biological quantification is necessary. Beak (1964) divides bottom invertebrates into three classes.

1. Group I

Includes those species known to be very tolerant of polluted conditions
and that frequently occur in very large numbers of individuals in polluted
conditions: Tubifex sp. is an example.

2. Group II

Includes those species known to occur in both polluted and unpolluted
situations but that do not form the very large communities of single species
characteristic of seriously polluted situations. These include the midge
larvae (Chironimidae), certain crustaceans (Gammorus and Asellus), many
snails and clays, some worms, and some leeches.

3. Group III

Includes only those species not occuring in situations that are significantly polluted--that is, are intolerant of pollution. This group includes
mayflies (Ephemeroptera), stoneflies (Plecoptera), some beetle larvae,
some dragonfly larvae, and some larvae (Trichoptera).

The application of this quantification method depends on the following
conditions:

1. The survey must include some unpolluted as well as polluted stations in
the survey area.

2. The sampling is carried out by quantitative methods so that an estimate
of density of population as well as the number of species is obtained.

A biological score is then allotted to each sample or the samples of each
station on the following basis:

> Normal complement of Group III scores 3 points.
> Normal complement of Group II scores 2 points.
> Normal complement of Group I scores 1 point.

Thus a normal unpolluted sample will score 6 points.

The results of this scoring method can be represented either as a map
or a graph (Fig. 5). A typical survey on a Canadian river receiving pollution
from a number of sources is described. At zero miles on the horizontal

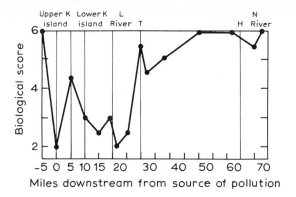

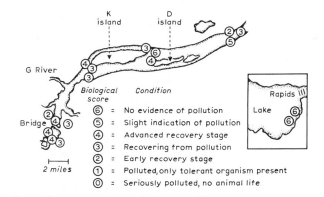

FIG. 5. Biological measurement of water pollution. From Beak (1964).

scale, a large city is located. Upstream the river is unpolluted. There is a large pulp and paper mill in the city between mile 0 and mile 5. A tributary feeds the river at mile 5. An integrated pulp and paper mill is situated between mile 10 and mile 15 and this causes a depression in the biological score. There is a gradual improvement until the L river is reached. There is another pulp and paper mill on this river with a noticeable effect on the biological score. The same depression occurs at two smaller towns downstream (T and H) where there are two further mills.

The biological score method may be compared with BOD, drop in dissolved O_2, and the total and suspended solids for a river (Fig. 6). The chemical results are an average of weekly measurements taken in 1958 and 1959 from May to October, while the biological score curve is a result of single surveys made in October of 1958 and 1959. A single biological survey in each year thus gives a fairly good comparison with seventeen chemical surveys in each year.

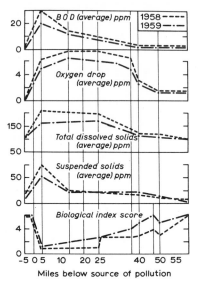

FIG. 6. Bioanalysis for water pollution. From Beak (1964).

VI. ANALYTICAL TECHNIQUE

Engineers are becoming increasingly aware of the need to apply good empirical solutions to the design of waste inputs into stream systems. Such techniques usually involve a concept of so many ft/sec of stream flow per 1000 of population. The experience of the engineer and his acquaintance with the characteristics of the stream in question are of particular significance in such an approach.

Recently, the construction of mathematical stream models based on the oxygen sag studies of Streeter (1958) has received particular attention from such workers as O'Connor (1966) at Manhattan College. The basis of the model is that the distribution of dissolved oxygen in a stream may be defined by the equation of continuity or mass balance in differential form:

$$\frac{dc}{dt} = \mu_{(x,t)} \frac{dc}{dx} \pm \Sigma_{(c,x,t)}$$

where c is the concentration of dissolved oxygen

      v is the velocity of flow in the x direction

      s is the sources and sinks of oxygen

The equation describes the temporal variations in oxygen in a stream in which there is a spatial distribution along the axis of flow. The concentration is assumed to be uniform over the cross-sectional area of the stream. The flow and the cross-sectional area from which the velocity is determined may be functions of space and time. The various sources and sinks of oxygen (s) may also be functions of time (t), space (x), and the concentration itself.

Sources of oxygen that are of significance in tributary flow are natural aeration and photosynthesis due to green plants. The sinks of dissolved oxygen include the utilization by the biochemical oxidation of organic matter, by benthal deposits, and by the respiration of aquatic plants. The organic utilization of oxygen may be divided into carbonaceous and nitrogenous components. The net gain or loss of dissolved oxygen in the system can thus be equated by inserting the total effects of sources against the effects of sinks. The problem with models of this type is that they require the quantification of biological parameters.

Although measurements of biological activity in stream waters are sufficiently sophisticated to be in use in such a model, the quantification of conditions in the sludge or stream bed can introduce significant errors into the system. The refinement of this tool allows the engineer to design wastewater treatment systems that relate actual effluent quality characteristics to quality resulting in a receiving stream.

VII. TECHNIQUES FOR CALCULATING LOADINGS OF STREAMS

The calculation of stream loading values is important not only in terms of conditions in the stream but also in terms of the effects these loadings have on the receiving basin. The sources of nutrients and solid wastes in a stream body are usually classified as (a) domestic wastes, (b) industrial wastes, and (c) other sources (mainly land runoff).

Measurements of domestic and industrial contributions are relatively easy because these values can be determined in the plant. The other sources (mainly comprised of land runoff) are usually determined by calculating the total loading where the stream runs into the lake by the formula:

concentration (ppm) X flow (ft^3/sec) X .9838 = loading (tons/year)

where 0.9838 is a constant to give the loading in ton/year.

The domestic and industrial contributions are then subtracted from the total loading to give the contribution from other sources:

total loading - domestic loading - industrial loading

= loading from other sources

One can assume that assimilation is nil because all factors in the nitrogen and phosphorus cycle are taken into account in the loading.

Thus nitrogen = Kjeldahl (organic nitrogen, includes ammonia)

+ nitrates (NO_3)

+ nitrates (NO_2)

Total nitrogen

phosphorus = total phosphates

As a typical example, the loadings for the Grand River flowing into Lake Erie are as follows:

mean flow = 1820 ft^3/sec (based on 1964-65 data)

suspended solids = 32,000 tons/year

dissolved solids = 416,000 tons/year

chlorides = 32,600 tons/year

P total = 1310 tons/year

136 municipal

458 industrial

716 other sources

N total = 1710 tons/year

1215 municipal

495 industrial

0 other

BOD = 3900 tons/year

(Concentrations are based on 1966-1967 OWRC data.)

The calculation of accurate stream loading figures for individual stream systems are a prerequisite for the calculation of mass balance loadings for large lakes such as Lake Erie and further analysis of loading relationships between large lakes that are connected as in the Great Lakes - St. Lawrence River system.

A good stream surveillance program will include systematic evaluations of chemical, biological, and physical changes. Background data should be collected for a year before a chemical plant or industry discharges into a river both above at the potential outfall and for at least 10 miles below the outfall. Such analyses will normally include about 20 individual tests. A producing plant should collect similar data on all wastes streams leaving the production site and should make similar investigations at least monthly or quarterly in the stream at points above and below the outfall.

Stream analysis studies will reduce the complexity of the stream from an art to a science even though one can expect continued difficulties in sampling and formulation of sophisticated mathematical models to quantify variables whether they be physical, chemical, or biological.

Streams will continue to be elusive because (Haney, 1958) they appear to be endowed with human characteristics:

"Streams can be willful, stubborn, secutive things and like people they seem to take great delight in confounding the experts."

REFERENCES

Beak, T. W. 1958. "The Effect of Liquid Wastes on Streams and Lakes."
Paper presented at the Symposium on Liquid Waste Disposal at the 41st
Annual Conference of the Chemical Institute of Canada, Toronto, May
28th.

Beak, T. W. 1964. "Biological Measurements of Water Pollution." Chem.
Eng. Papers, 60, No. 1.

Beeton, A. M. 1965. "Eutrophication of the St. Lawrence Great Lakes."
Limnology Oceanography, 10, 240-254.

Brinkhurst, R. O. 1966. "Detection and Assessment of Water Pollution
Using Oligochaete Worms (Parts 1 and 11)." Water Sewage Works
(Oct. and Nov.)

Browzin, B. S. 1962. "On Classification of Rivers in the Great Lakes - St.
Lawrence Basin." Pub. #9, Great Lakes Research Division Institute of
Science and Technology, The University of Michigan.

Dappert, A. F. 1958. "The Use of Stream Data in Administration of Pollution Abatement Programs." From Oxygen Relationships in Streams
(Proceedings of Seminar sponsored by the Water Supply and Water Pollution Control Program, U.S. Dept. of Health Publication.)

Dawson, R. N. 1962. "The Grand River - A Study of Stream Pollution and
Water Supply." Thesis for the Degree of Master of Applied Science,
University of Toronto.

Doudoroff, P. 1951. "Biological Observations and Toxicity Bioassays in the
Control of Industrial Waste Disposal." Proceedings 6th Industrial Waste
Conference Purdue Eng. Bulletin #76, pp. 88-104.

Haney, P. D. and Schmidt, J. 1958. "Representative Sampling and Analytical Methods in Stream Studies from Oxygen Relationships in Streams."
Proceedings of Seminar sponsored by the Water Supply and Water Pollution Control Program. U.S. Dept. of Health Education and Welfare.

Hoskins, J. K. 1938. "Planning the Organization and Conduct of Stream
Pollution Surveys." Public Health Research, 53, 729.

Hynes, H. B. W. 1963. The Biology of Polluted Waters. Liverpool University Press.

Johnson, M. G. 1964. "Review of Some of the Literature on the Biological Assessment of Polluted Waters." OWRS Publication.

Johnson, M. G. 1966. "A report on the scope and organization of the biological survey program." Publ. Biology Branch, Ontario Water Resources Comm.

O'Connor, D. J. and DiToro, 1966. "An Analysis of the Dissolved Oxygen Variation in a Flowing Stream from Stream and Estuarine Analysis." Manhattan College Summer Institute in Water Pollution Control, New York (pp. 1-11).

Ontario Water Resources Commission. 1967. "Detection and Measurement of Stream Pollution." Proceedings of the 4th Annual Technical Seminar (Professional Engineers Group).

Patrick, Ruth, 1953. "Biological Phases of Stream Pollution." Proc. Penn. Acad. Sci., 27, 33-36.

Patrick, Ruth, 1956. "Diatoms as Indicators of Changes in Environmental Conditions." Trans. Sem. on Biological Problems in Water Pollution, U.S. Public Health Service (pp. 71-83).

Streeter, H. W. 1958. "The Oxygen Sag and Dissolved Oxygen Relationships in Streams from Oxygen Relationships in Streams." U.S. Department of Health Education and Welfare (pp. 25-32).

Thomas, J. F. J. 1953. "Scope Procedures and Interpretation of Survey Studies." Water Pollution Report, 1. Queen's Printer, Ottawa.

Welch, Paul S. 1948. Limnological Method. Blakiston, Philadelphia.

Chapter 6

CONTROL OF AQUATIC PLANTS AND ALGAE

Water plant control of lakes, ponds, or water areas requires:

1. Identification of the nuisance plant or organism.
2. Calculations of the area or volume of water to be treated.
3. Review of the choice of chemicals, and determination of costs.
4. Possible application to obtain a permit (if a permit is necessary) for treatment
5. Review of general suggestions for applying herbicides

Many chemicals have been marketed in recent years that are effective in controlling algae and other aquatic plants. Certain chemicals are useful in fish reclamation work and others have been used for the control of insects, snails (swimmers' itch), and leeches.

Section 28b, subsection 1 of the Ontario Water Resources Commission Act provides that "No person shall add any substance to the water of any well, lake, river, pond, spring, stream, reservoir or other water or watercourse for the purpose of killing or affecting plants, snails, insects, fish or other living matter or thing therein without a permit issued by the Commission." Provision has been made for regulations under the Act to exempt persons, substances, or any quantity or concentration thereof from the application of Section 28b subsection 1. Two regulations have been enacted which, without the authority of a permit, enable:

1. A person to apply any substance to a pond which is wholly enclosed by property and has no outflow; and

2. A person to apply a substance to a drainage ditch for the purpose of controlling emergent aquatic plants, if he uses 2,4-D or 2,4,5-T, or a mixture of the two, at a rate not exceeding 3 lb acid equivalent per acre, or any other substance approved for controlling emergent vegetation under the Pest Control Products Act of Canada when applied according to the directions provided with the substance.

In all other cases where substances are to be used for controlling or affecting life in water, a permit must be obtained from the Commission.

Legislation related to the control of aquatic nuisances ensures that there will be no unreasonable infringements on the rights of other water users and that substances will not be used that are toxic to humans, fish, domestic animals, and wildlife. It also ensures that stable compounds will not be used that might tend to accumulate in water, thus posing a threat to potable water supplies.

To secure a permit for applying a chemical or other substances to control nuisance conditions in any area of water, an individual or commercial agency must submit pertinent information on an official application form. Thus, the nature of a project and possible consequences arising from it may be satisfactorily evaluated. These application forms may be obtained in writing the proper governmental authority. An application should be submitted well in advance of the time that the chemical is to be applied. While every effort is made to process applications as quickly as possible, delays of up to a month are encountered, particularly if the site has to be visited.

The acquisition of a permit does not divest any individual or commercial applicator of the responsibility for any undesirable consequences arising from a treatment. Anyone applying any substance without the authority of a permit, or who violates the terms and conditions provided in a permit, is guilty of an offense and liable to a fine, which in Ontario will not exceed $500.

I. AQUATIC VEGETATION

High aquatic plants and algae are important in maintaining a balanced aquatic environment. However, depending on the uses made of the water, there may be situations where their presence is undesirable.

On the positive side, in addition to maintaining an oxygen balance essential to fish life, water plants provide a suitable environment for the production of aquatic invertebrate organisms, which serve as food for fish. They also contribute to keeping water temperatures at the low levels essential to certain species of fish and provide shade and protection for young game fish and forage fish species. Finally, numerous aquatic plants are utilized for food and/or protection by many species of waterfowl.

On the other hand, ponds and lakes may become unsightly because of the presence of dense mats of decomposing surface-type algae. Recreational uses such as fishing, swimming, or boating may be impaired by heavy accumulations of algae or thick growths of high aquatic plants. Decaying masses of vegetation may cause water to become less palatable to humans or to domestic livestock. Finally, winter-kills of fish may result from oxygen depletion in the water caused by a decomposition of plants under the ice during certain winter seasons.

Certainly, a careful assessment of the various usages and relative values of the presence or absence of aquatic plants in a particular situation should be made before any control project is undertaken.

II. CONTROL OF AQUATIC PLANTS

Control of aquatic plants may be achieved by either mechanical or chemical means. Simple raking and chain-dragging operations to control submergent species may produce temporary results, but the plants soon re-establish themselves. Underwater mowing and dredging machines are used successfully to keep channels open for boating and to provide access to docks and good fishing areas; however, acquiring and maintaining equipment of this kind involves considerable expense. Shoreline emergent plants may be hand-pulled or cut with a scythe, and these methods may be practical if the area involved is not too large.

Chemical methods of control are the most practical, considering the ease with which they can be applied. However, the herbicides and algicides currently available generally provide control for only a single season. A satisfactory algicide or herbicide must kill the plant or plants causing a nuisance, should not affect fish or other aquatic life, and should be reasonable in cost.

At the present time there is no one chemical that will adequately control all species of algae and other aquatic plants. In selecting a particular chemical, the species for which control is desired must be considered, as well as the temperature and chemical properties of the water.

III. TYPES OF AQUATIC PLANTS

Aquatic plants may be divided into three categories, as follows:

1. Submerged rooted aquatics.

2. Emergent plants -- may have upright leaves or leaves that float on the surface of the water.

3. Algae -- color the water green or brown, or appear as "pond scum."

Some herbicides control several to many species within either the submergent or emergent groups; others control plants in both these categories with varying effectiveness. Some chemicals will control higher aquatic plants and algae as well.

IV. OPTIMAL CONDITIONS FOR TREATMENT

Algae and rooted submergent plants should be treated during the spring or early summer while the plants are developing rapidly and before they reach nuisance proportions. During this period, the chemical will provide more effective control of the plants and there will be less likelihood that fish mortalities will be caused by oxygen depletion, which can result from the decomposition of a large plant mass. Algicides and herbicides are generally more effective in warmer water and better control will be achieved if the water temperature is over $65°F(18°C)$.

Control of emergent vegetation should be undertaken about the time of flower or seed-head formation, on days that are calm and sunny. Windy weather increases the hazard to the person applying the chemical and to nearby valuable plants. If rain falls shortly after a spray is applied it will wash the chemical off the plants, thereby reducing the effectiveness of the treatment.

V. CALCULATION OF WATER VOLUMES AND DOSAGE RATES

When control of submerged plants is attempted using liquid-type chemicals, it may be essential to know the volume of water present in the area to be treated. The surface area must be calculated and the average depth should be determined by adequate sounding.

To determine the volume and total weight of water in the area to be treated, the following procedure is used:

Length x width x average depth = volume in ft^3

Volume in ft^3 x 62.4 lb = total weight of water (1 ft^3 of water weighs 62.4 lb)

Note: If area is measured in acres, 1 acre = 43,560 ft^2

Effective concentrations of a herbicide or algicide are expressed as so many parts of active chemical per million parts of water. Many chemicals are not sold in a pure state but are marketed as water or oil-base solutions or are impregnated in granules of inert materials. Therefore it is imperative to know the percentage of pure active ingredient(s) in the product to be used, and this information is provided on the label of the container in which the chemical is sold.

Recommended concentrations, expressed as ppm (1 lb of active chemical per million lb of water = 1 ppm), are usually provided by the manufacturers or suppliers of control products.

The following formula is used to calculate the number of pounds of active chemical required:

Total weight of water in lbs. x recommended concentration of active chemical in ppm.

1,000,000

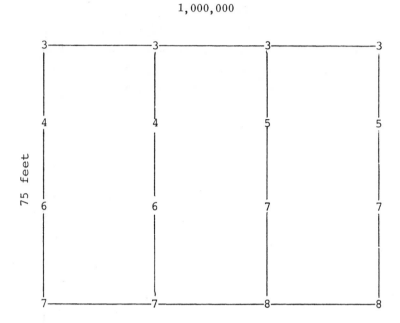

Scale 1" = 50 feet

Example: Calculate the number of pounds of 20% active chemical required to treat the plot illustrated, at a concentration of 2 ppm.

Total area = 7500 ft^2

Average depth = 5.4 ft

Total water volume = 7500 x 5.4 = 40,500 ft^3

Total weight of water = 40,500 x 62.4 = 2,527,200 lb

Applying the formula:

$$\frac{\text{Total weight of water x recommended concentration of active chemical in ppm}}{1,000,000}$$

$$\frac{2,527,200 \times 2}{1,000,000}$$

= 2.5 x 2 or 5 lb of active chemical

Therefore, since this particular material is only 20% active, 25 lb of the commercial product would be required.

Figure 1 illustrates dosage rates in determining the total weight of chemical required to treat a pond at various concentrations. The graph does not cover all situations, however, and it may be necessary for the reader to make his own calculations.

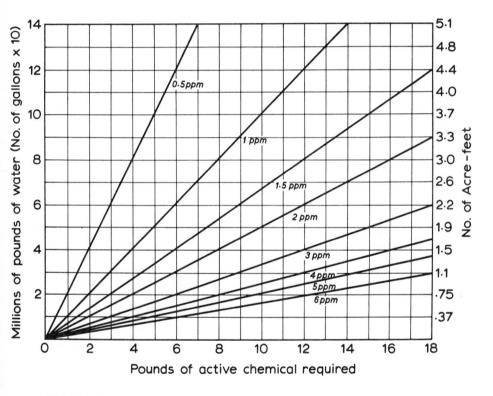

FIG. 1. Dosage rate curves

Producers of some granular herbicidal products recommend application rates based on pounds per acre, regardless of the depth of water present. Since the granules drop to the bottom of the lake or pond, recommendations of this nature are based on the premise that the chemical is absorbed through the roots of the plants.

Where emergent species are treated with contact sprays, the manufacturer's instructions should be followed concerning the percentage solution

required to provide effective control. Since the chemical is sprayed directly
on the plants, it is not necessary to calculate the weight of water in the pond.
It should be remembered that the recommended percentage solution may be
based on the active ingredient(s) rather than the commercial formulation.

VI. GENERAL SUGGESTIONS CONCERNING USE OF HERBICIDES AND ALGICIDES

Before any chemical control measures are undertaken, all riparian
owners adjacent to and in the general vicinity of the treatment area should
be notified. Due consideration must be given to any objections voiced by
other parties who may utilize water from the surrounding area for drinking,
swimming, fishing, watering domestic animals, and irrigation. Use of
treated water following any application should be restricted in accordance
with directions provided by the manufacturer or supplier of the chemical.

Where fish are present and there is a heavy growth of algae or aquatic
plants, the entire pond or lake should not be treated at one time. As mentioned previously, decomposition of a large plant mass can lead to depletion
of the dissolved oxygen supply so that the fish will suffocate. Under such
circumstances, several sectional applications should be undertaken, spaced
about a week apart.

Where algicides or herbicides are actually mixed with or distributed
throughout the water, it is imperative that an even distribution of the chemical be effected throughout the area to be treated. If localized high concentrations develop, destruction of fish and other aquatic life may result and
spotty control of the plants will be achieved. The amount of chemical applied
should be in proportion to the depth of water throughout the area to be
treated.

Since certain herbicides and algicides must be handled carefully because
of their toxic properties and sometimes corrosive nature, the specifications
for use which are provided by the manufacturer or distributor should be
followed closely.

A. Methods of Application

For larger projects, where submergent plants are being treated, the use of a power-driven pump mounted in a boat is desirable. The pump should be fitted with a dual intake so that the chemical can be diluted by water taken in through a hose attached to a foot valve suspended over the side of the boat. The diluted chemical should be injected underwater through a distribution boom fitted with weighted trailing hoses. When using a boat on smaller areas, liquid compounds that are not dangerous to handle may be diluted to a 5% solution and added in a regulated flow to the slipstream of the outboard motor. The action of the propellor will tend to disperse the chemical.

Granular products may be applied using a rotary-type seeder, with care being taken to ensure an even distribution of the material. Copper sulfate, which is often used effectively for control of algae, is a crystalline material that should not be breathed or allowed to get in the eyes. This chemical should be dissolved and applied as a liquid using spray equipment or placed in a burlap bag and dragged behind a power boat. The speed at which the boat is operated should be related to the rapidity at which the dissolved chemical passes from the bag into the water and the deeper water should receive proportionately more chemical.

Emergent aquatic plants may be sprayed using a backpack sprayer of the kind used to spray weeds in lawns. All of the exposed foliage should be thoroughly wetted and the addition of household detergent to the spray tank will improve the wetting action. Half a cup of liquid detergent will make 50 gal of spray.

Orchard-type guns may be used for spraying emergents on larger areas, and aircraft have been used to apply chemicals to cattails and other emergent plants where the size of the area sprayed has warranted the use of this method.

B. Information on Chemicals

Information concerning specific herbicides and algicides should be obtained from the manufacturer. The Advisory Committee on Herbicides for Ontario publishes recommendations each year in Publication 75 of the Ontario Department of Agriculture. Pertinent extracts from this publication

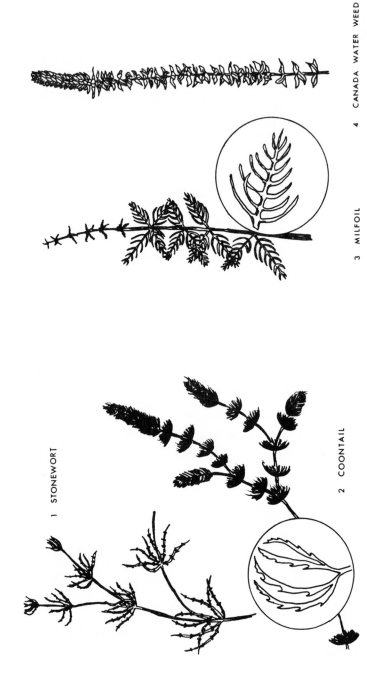

FIG. 2. Submergent and emergent aquatic plants in lakes in Canada and the northern United States. Submergent plants: (1) Stonewort (Chara sp. of algae), (2) Coontail (Ceratophyllum sp.), (3) Milfoil (Myriophyllum sp.), (4) Canada water weed (Anacharis canadensis or Elodea). Drawing also shows expanded view of parts of leaf (scale not shown). (Sheet 1 of 4)

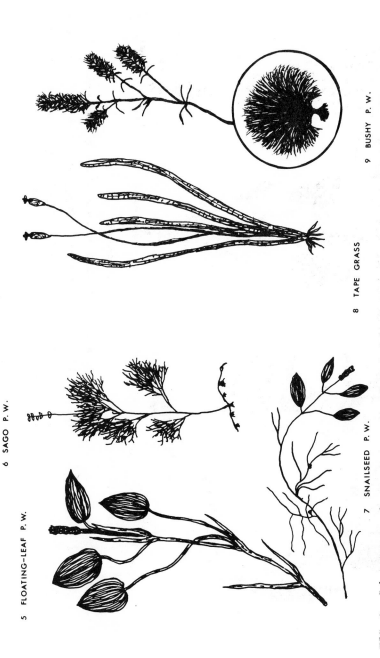

5 FLOATING-LEAF P. W.

6 SAGO P. W.

7 SNAILSEED P. W.

8 TAPE GRASS

9 BUSHY P. W.

FIG. 2. Submergent and emergent aquatic plants in lakes in Canada and the northern United States. Submergent plants: (5) Floating-leaf pond weed (Potamogeton natans), (6) Sago pond weed (Potamogeton pectinatus), (7) Snailseed pondweed (Potamogeton diversifolius), (8) Tape grass (Vallisneria americana), (9) Bushy pondweed (Najas flexilis). Drawing also shows expanded view of parts of leaf (scale not shown). (Sheet 2 of 4)

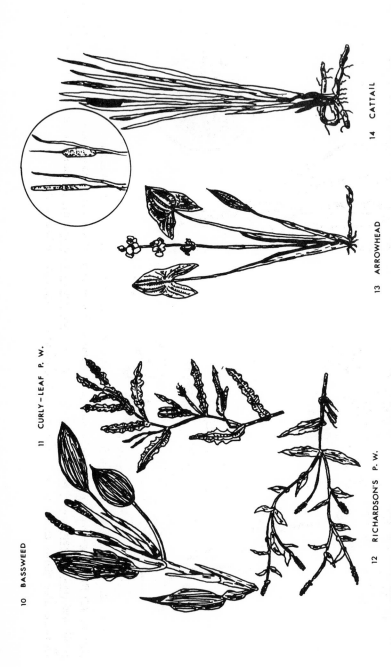

10 BASSWEED

11 CURLY-LEAF P. W.

12 RICHARDSON'S P. W.

13 ARROWHEAD

14 CATTAIL

FIG. 2. Submergent and emergent aquatic plants in lakes in Canada and the northern United States. Submergent plants: (10) Bassweed (Potamogeton amplifolius) , (11) Curly-leaf pondweed (Potamogeton cripus) , (12) Richardson's pondweed (Potamogeton richardsonii). Emergent Aquatics: (13) Arrowhead (Sagittaria latifolia), (14) Cattail (Typha latifolia). Drawing also shows expanded view of parts of leaf (scale not shown). (Sheet 3 of 4)

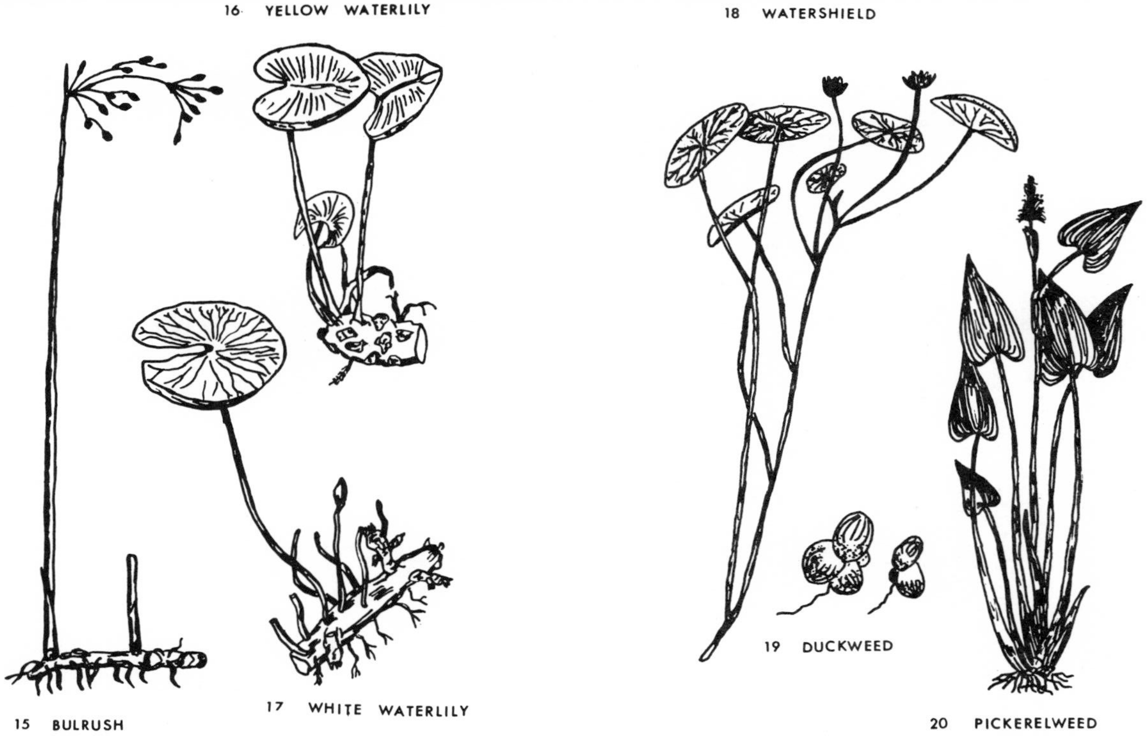

15 BULRUSH

16 YELLOW WATERLILY

17 WHITE WATERLILY

18 WATERSHIELD

19 DUCKWEED

20 PICKERELWEED

FIG. 2. Submergent and emergent aquatic plants in lakes in Canada and the northern United States. Emergent plants: (15) Bulrush (Scirpus validus), (16) White waterlily (Nymphaea tuberosa), (17) Yellow waterlily (Nuphar advena), (18) Watershield (Brasenia schreberi), (19) Duckweed (Lemna minor), and (20) Pickerelweed (Pontederia cordata). Drawing also shows expanded view of parts of leaf (scale not shown). (Sheet 4 of 4)

and a list of suppliers of chemicals used for aquatic plant control are available upon request from the Ontario Water Resources Commission of appropriate commission.

VII. PLANT IDENTIFICATION

The following pages illustrate some of the more common submergent and emergent aquatic plants that cause problems (Fig. 2) in lakes and farm ponds throughout the northern U.S. and Canada. The labels of containers in which approved herbicides are marketed indicate which aquatic plants may be satisfactorily controlled.

REFERENCE

Aquatic Plant and Algae Control, Ontario Water Resources Commission, 801 Bay Street, Toronto 5, Ontario.

Chapter 7

ACTIVATED SLUDGE

I. ESSENTIALS OF ACTIVATED SLUDGE

This is a natural biological process that developed as a result of man's actions rather than through invention. It was found that when sewage or any organic waste was aerated for a sufficient time, the organic matter content was reduced and a flocculant sludge formed. Recycling sludge and developing a continuous process have been the primary new contributions to the process.

Activated sludge is an aerobic biochemical process by which microbes degrade organic waste producing carbon dioxide, water, and energy. Many individuals have shown the involvement of microbes. When the oxidation is not complete, intermediate organic compounds are formed. Some of the energy produced is utilized to generate additional cells. These additional cells contain a lot of slime often described as a zoogleal mass. This complex zoogleal mass is activated sludge. When conditioned properly, sludge aides in several distinct processes in the activated sludge process.

Before analyzing these processes, the basic activated sludge process should be examined (Fig. 1). A basic unit consists of an aeration basin and an thickener. The aeration basin is the heart of the process; however, the sludge produced must be settled out and part of the sludge recycled to be a true activated sludge process. The influent waste is usually diluted about one-fifth with recycle sludge. The sludge concentration in the aeration basin should be maintained between 1500-2500 mg/liter. The resulting BOD reduction is about 90-95%.

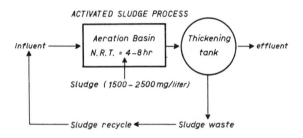

FIG. 1. Flow chart of activated sludge process.

In a standard aeration basin the nominal retention time is 4-8 hr. Another rational criterion for operation is the volumetric BOD loading. Greeley (1945) recommends 30 lb. BOD/1000 ft^3 of aeration capacity.

There are many variations in the activated sludge process. All variations were developed to resolve special problems encountered with the standard process. Before examining these variations a closer analysis must be made of the bioprocesses involved. In mixing the recycle sludge with the influent waste, the first reactions are one of adsorption and absorption of both soluble and insoluble organics. Some inorganics are likewise removed but this is not a primary function of the activated sludge process. A well-conditioned activated sludge possesses high sorptive capacity.

The second process of major importance is oxidative catalysis. The long retention time in the aeration basin permits the complex microbial population in the sludge to oxidize a major portion of the organic waste to CO_2 and water, thus reducing the BOD. The level of dissolved oxygen in the aeration basin and the continuous supply of oxygen become a primary concern to the engineer. Air is usually supplied by sparging using surface aeration with mechanical agitators. The sludge waste passing from the aeration basin to the thickener should be low in BOD. The thickener should be sized to permit adequate settling and removal of sludge. Bulking is encountered at times. This is a condition where the activated sludge is light and fluffy and even fragile and will not settle. The sludge then flakes over the separating weirs and out with the overflow effluent waste. Bulking is discussed later.

Microbes that do not settle with the floc are also continually washed out of the system. Thus the activated sludge process is a selective process, continually enriching for microbes, which cause flocculation and settle. Those microbes that produce flocculating agents that cause them to settle out in the thickener gradually become predominant in the system. Selective conditions are for BOD removal, settlability, and to a lesser extent, compactability. Since the flocs are light, having a density close to water, turbulent conditions have to be reduced to accomplish settling.

Although the activated sludge process has been in operation over 50 years, there is still inadequate information for developing exacting design data. Pilot plant studies are almost always needed before a design change can be introduced. Parameters that need investigation for proper design are mean BOD of influent feed and volume, pre-aeration and grit removal requirements, aeration requirements, rate of sludge production, thickener requirements, and proportion of sludge recycle and sludge wastage requirements. The type of mixing in the aeration basin must be considered because this has a direct bearing on efficiency and sizing. In addition, provisions may have to be made to incinerate waste sludge or to consider other means of disposal. Chlorination and tertiary treatment to remove phosphate from the effluent may be required.

II. BIOLOGY AND THE ENERGY CHAIN

The organism population in activated sludge is so complex that it can only be defined in terms of broad ecological groups of organisms. Those commonly found are bacteria, protozoans, metazoans, and fungi. If sunlight is adequate, algae are also found. Contaminating organic matter may be oxidized partially by one group of microbes and the intermediates formed used as energy sources by a second group of microbes. In an efficient process the organic matter is oxidized to CO_2 and water with the production of new cellular material. Because the process is biological, some of the important microbes found are protozoa such as

1. Sarcodina -- an amoeboid cell possessing psuedopodia which absorbs soluble organics through membranous cell walls and engulfs particulates.

2. Mastigophora -- divided into holophytic types, e.g., Euglena and Chilomonas, and holozoic types, e.g., Oikomonas.

3. Ciliata -- ciliated free-swimming organisms of Paramecium and Tetrahymena and the stalked types of Vorticella, Glaucoma, and Espistyla.

A general distribution curve for organisms at various stages in a conventional activated sludge is shown in Fig. 2(a) (McKinney and Gram, 1956; Cooke and Ludzack, 1958; Stanier et al., 1963; Schenk and Howes, 1963). As the bacteria begin to increase, their presence causes an increase in the number of Sarcodina, Vorticella, Paramecium, and Mastigophora.

Sketches of protozoa commonly found in sludge are shown in Fig. 2(b). The holophytic flagellates cannot compete with the bacteria for the soluble organic matter. The bacteria are smaller and possess a much greater sorptive capacity per cell. As the COD-to-sludge ratio decreases, the holozoic Mastigophora and later the face-swimming Ciliata act as primary predators, increasing in numbers by feeding on bacteria. With continued aeration, further organic materials are removed and stalked ciliates make an appearance along with secondary predators known as Rotifers. The ciliates are observed until flocs are formed, which appears to inhibit their growth. The abundance of Rotifers is a good indicator of a low BOD waste since these small animals do not survive in highly polluted water.

A. Temperature

Temperature has a controlling effect upon all forms of biological activity. Individual species of organisms have definite temperature ranges for growth and metabolism. In a mixed heterotrophic population the general temperature range is 15-40°C. Microbes may be classified by their temperature requirements into psychrophiles (<15°C), mesophiles (15-45°C), and thermophiles (>45°C). Comparative studies of each of these groups upon defined substrates in the activated sludge process are needed. Such data may show that biopopulations found in sludge in midwinter have acclimated to

the temperature change and are oxidizing substrates at the same rate as
mesophilic populations. In dealing with psychrophiles Ingraham and Stokes
(1959) suggest that these microbes should be defined in terms of their minimal temperature for growth and not in terms of temperature optima. Utilizing this criterion a psychrophile is defined as an organism that grows well
for one week at $0°C$. This concept has not been extended to either mesophiles
or thermophiles.

Individuals dealing with organisms should realize that microbes grown
at temperatures suboptimal for growth possess characteristics quite different from those grown under optimal temperature conditions (Ingraham and
Stokes, 1959; Cairns, 1956; Ludzack et al., 1961). Some of the more notable
characteristics are (a) larger unicellular volumes, (b) greater proportions of
ribonucleic acids and proteins, (c) greater pigment concentrations, (d) tendency to produce greater amounts of polysaccharides, (e) lower rates of
metabolism, and (f) increased proportions of unsaturated fatty acids (up to
50%) in the lipids synthesized.

The physiological role of unsaturated lipids has been shown experimentally to have an affect on membrane permeability. A cold shock to microorganisms results in a release of certain low-molecular-weight constituents
(Rose, 1968). Psychrophiles, with larger proportions of unsaturated lipids,
do not show this susceptibility, nor do microorganisms in general exhibit the
same degree of shock for a rise in temperature (Ludzack et al., 1961).
Membrane permeability is not the only limitation to cellular activity (Stokes
and Larkin, 1968). Experiments with cell free extracts, where substrate
permeability was not involved, showed that psychrophiles functioned better
oxidatively at low temperatures.

Many regulatory processes are known to be abnormally sensitive to
temperature. In general, the apoenzyme, or protein portion of an enzyme,
is more thermolabile than the prosthetic coenzyme. The organic nature and
smaller size of the latter enable it to diffuse through membranes more
readily, wherein it can undergo reactions with other apoenzymes. It is the
ability to alter enzyme structure that imparts to the microbe the ability to
adapt to new substrates and environmental conditions.

Data on the effects of temperature on the operation of an activated
sludge process are at times inconsistant. Cairns (1956) observed a change

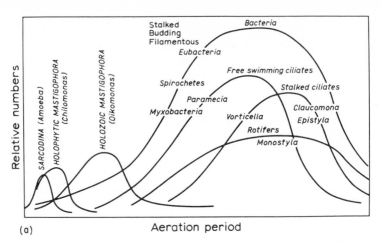

FIG. 2(a). Microbial distribution -- conventional activated sludge
(McKinney et al., (1956); Stanier et al., 1963; Schenk et al., 1963).

in microorganism species and distribution with a change in temperature.
Porges et al. (1955) found that a decrease in temperature from 30 to 2°C had
much greater affect on oxygen utilization than on oxygen demand. This
agrees with Ludzack et al. (1961) who observed that lower temperatures result in a decrease in catabolism, giving rise to greater solids gain per unit
weight of feed.

Further experiments showed that BOD and COD removal was about 10%
higher at 30°C than at 5°C when the influent was treated at both temperatures.
At low temperatures flocculation was substantially inferior to that at 30°C
until exogenous sources of energy become limiting, and catabolism becomes
significant in total respiration. The acclimation to shock changes in temperature took about five times longer at 5°C than at 30°C. Soluble substances
are metabolized more slowly at low temperatures and contribute to poor
flocculation. Sphaerotilus apparently becomes more competitive at 5°C and
nitrification decreases. Addition of excess inorganic nitrogen at 5°C is
detrimental. At the lower temperatures foaming becomes appreciable and
the sludge volatile percentages are about 5% higher. Most authorities do not
design differently for summer and winter treatment. This may not be true
as more data are developed.

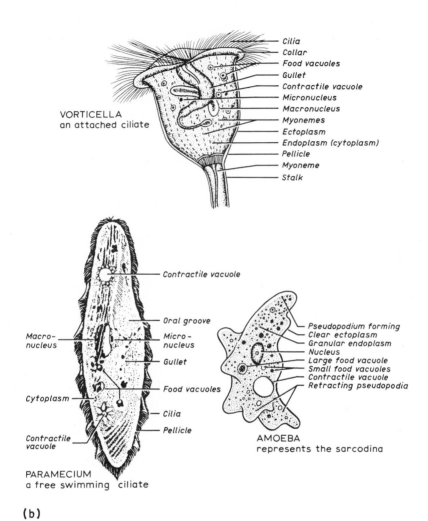

(b)

FIG. 2(b). Important Protozoa found in activated sludge.

B. Concentration and Type of Feed

Feeds can be classified as soluble and particulate. Soluble wastes, if biodegradable, are used directly by the microorganism. Insoluble substrates are those that will not dissolve in the aqueous medium. In the process of adsorption they generally require shorter contact times (Norman and Busch, 1966). Although absorbed more readily, they require solubilization before being consumed by the microorganisms. The latter process is carried out by extracellular enzymes, which hydrolyze the organic solids.

In the absence of available food, cells consume internally stored products, which can lead to serious lag times when re-exposed to new feed. If the stabilization time is limiting, high sludge-to-feed ratios should be avoided. On the other hand, Ford and Eckenfelder (1967) have shown that high organic loading can lead to an excessive filamentous-type growth. Short periods of exposure to anaerobic conditions have been shown to kill the majority of filamentous organisms and yield better settling, without affecting the assimilative capacity of the system (Ford and Eckenfelder, 1967). As mentioned earlier, high organic loading leads to proportionally less catabolism and larger solids gain per unit weight of feed.

Domestic wastes are known to consist largely of colloidal and suspended organic matter, whereas large portions of industrial wastes are soluble.

Rao and Gaudy (1966) conducted long-term experiments (3 months after acclimation) on activated sludge units. Although they maintained constant light, temperature, pH, and feed-to-sludge ratio, they experienced a cycling of microorganism population strongly related to the substrate removal rates. They reported a proportional decrease in oxygen utilization with increasing initial sludge concentration, which agreed with Ford and Eckenfelder (1967).

III. KINETICS

A. General Information

In analyzing the overall activated sludge process one has to be concerned with (a) sludge production, (b) enzyme kinetics, (c) aeration, and (d) the premixing. Settling rates should be examined; however, they are treated under coagulation and settling.

Sludge growth is comparable to cell growth, which can be described by a classical growth curve (Fig. 3). Microbial cells often divide by binary fission, with each cell producing two new cells and the cell number increasing in a geometric progression, i.e. exponentially. Pure cultures possess

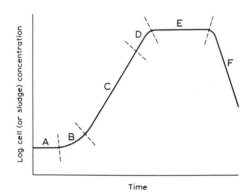

FIG. 3. Cell (or sludge) production curve in a batch tank; (A) lag phase; (B) accelerating growth phase; (C) log phase; (D) transition phase; (E) stationary growth phase; (F) declining growth phase.

characteristic growth rates when grown on defined substrates. The rate of growth of a culture is directly proportional to the number of cells present at that time (or to the weight of cellular tissue at that time).

$$dN/dt = kN \qquad (1)$$

where N = number of cells

k = growth rate constant

Integrating this expression yields

$$N = N_o e^{kt} \qquad (2)$$

where N_o = cells at zero time

N = cells at time t

k = growth rate constant

Solving for k:

$$k = \frac{\ln (N/N^{o})}{t} \tag{3}$$

Here k represents the rate at which the natural logarithm of cell number increases with time. Usually a plot of the "increase in cell number" against time gives a straight line whose slope is k (Fig. 4). Growth rates vary with the type of culture and the substrate. They are always determined in excess

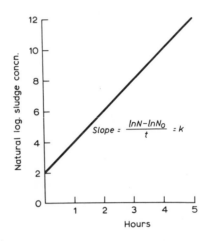

FIG. 4. A plot of the logarithm of cell (sludge) concentration versus time.

substrate. The time for the cell population to double is called the generation time. Growth rates are often reported in generations per hour. In an actively growing culture the number of cells N increases with each generation:

g	N
0	1
1	2
2	4
3	8
4	16
5	32

This geometric progression has the form

$$N = N_0 2^g \qquad (4)$$

Combining eqs. (2) and (4):

$$N_0 2^g = N_0 e^{kt} \qquad (5)$$

Rearranging and cancelling N_0:

$$2^g = e^{kt} \text{ or } g\ln 2 = kt \text{ or } \frac{g}{t} = \frac{k}{\ln 2}$$

This last equation relates the generation time to the growth constant k. Because one is dealing with a very complex microbial generation,population time really has no application in the activated sludge process; however the rate of sludge growth(s) follows the same exponential function described for cell growth. Thus S could be substituted for N.

Another analogy between cell growth and sludge growth can also be made. Sludge growth in a batch vessel follows a classical cell growth curve (Fig. 4). The growth curve can be divided into at least 6 phases. These are A: lag, B: acceleration, C: exponential (or log), D: retardation, E: maximum stationary, and F: declining growth. The changes in growth rate,dN/dt, during each phase are summarized in Table 1. Note that the exponential phase is constant and that lag and maximal stationary phase are zero.

A logarithmic plot of cell (in sludge) growth versus time usually gives a linear plot (Fig. 5). The slope k is a constant characteristic of the system under study.

TABLE 1

DIVISIONS OF CLASSICAL GROWTH CURVE IN BIOSPHERES

Section of curve	Phase	Growth rate
A	Lag	Zero
B	Acceleration	Increasing
C	Exponential	Constant
D	Retardation	Decreasing
E	Maximum stationary	Zero
F	Decline	Negative (death)

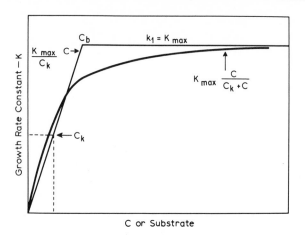

FIG. 5. One and two phase growth curve constants (Wilson, 1963).

B. Kinetics of Activated Sludge

The activated sludge process must be analyzed in terms of microbial growth or sludge production from organic substrates (or wastes). Fundamental to understanding this process is a knowledge of microbial growth. Whether analyzing a single culture or growth of a mixed culture on a waste organic compound, batch growth will be expressed in terms of the main growth phases discussed earlier, i.e. lag, logarithmic, stationary, and declining. Logarithmic growth is important in activated sludge or any bioprocess because activities occurring in this stage are exponential. It is also a stage where the properties of the cells are most constant.

There are problems in applying growth kinetics to activated sludge, and details of these problems must be examined. First, growth kinetics are substrate-dependent. Whereas most kinetics studies are completed in systems with abundant substrate, this is not the general situation in activated sludge. Many sewage supplies average around 300 ppm of BOD, which is a limiting amount of substrate in most microbial processes. Also, the assumption of an overall velocity during steady state places some operational limitations on the use of the Monod (1949) equation in defining the overall rate of an activated sludge process. This is true because this rate is an average rate,

dependent on many consecutive reversible reactions and the equilibrium
constant of each reaction. This is further complicated by mixed substrates,
mixed cultures, pH, aeration, and mixing. Thus, kinetic formulas must
include a large number of assumptions.

The equation used by Monod (1949) and applied by others in activated
sludge (Garrett et al., 1952; Hoover and Porges, 1952) describes a general
hyperbolic function. Equations can be written for the rate change relating to
cell mass (i.e., sludge) X; product (i.e., CO_2)/P; and substrate (i.e.,
BOD/S. In a batch system these rates would be expressed by

Rate of growth: $$\mu = \frac{1}{x} \frac{dx}{dt}$$

Rate of product formation: $$\mu = \frac{1}{x} \frac{dP}{dt}$$

Rate substrate utilization: $$\mu = \frac{1}{x} \frac{-dS}{dt}$$

In a batch reactor the rate is a function of time and the maximum value
of specific growth rate, referred to as μmax, is obtained in the exponential
phase of growth. μmax is normally calculated under conditions where substrates are in excess:

$$\mu max = \frac{1}{x} \frac{dx}{dt} = \frac{d \ln X}{X}$$

Solving this differential we obtain

$$X = X_o e^{\mu max (t-t_o)}$$

where t_o = time at threshold of logarithmic growth phase

X_o = microbial cell concentration when $t = t_o$

There are two main methods of representing the kinetics of the biochemical reactions occurring in the activated sludge process. One proposed by
Monod (1949) and modified by Wilson (1963) and the other proposed by Garrett
and Sawyer (1952). The single-phase theory of Monod estimates the growth
rate of a pure bacterial culture undergoing successive divisions. During the

exponential phase, the growth rate is constant and reaches a maximum value.

The empirical equation is given by the hyperbolic function

$$K = \frac{1}{S} \cdot \frac{ds}{dt} = K_{max} \frac{C}{C_k + C} \qquad (6)$$

where K = the exponential growth rate

C   = concentration of limiting nutrient at time t

K_{max} = the maximum value of the growth rate constant at infinitely large substrate concentration

C_k = Michaelis-Menten constant, which is defined as the concentration of the limiting nutrient at which the growth rate is $0.5\ k_{max}$

S   = the concentration of microorganisms

Those interested in kinetic models used in design should consult many of the references directly. Gram (1956) has applied reaction kinetics to aerobic biological processes in which activated sludge was included. Popel (1962) used both laboratory data and plants in Germany in developing reaction kinetics. Aiba et al. (1965) provide detailed information on the "scale up" of aerobic processes encountered in biochemical engineering. Simple formulae have been developed for calculating sludge accumulation (SA) from sludge loading data (SL) (Oxford et al., 1963):

SA = 0.313 log SL + 0.593

where SA = sludge accumulation (lb volatile solids/lb BOD removed)
      SL = sludge loading (lb BOD/day/lb of volatile solids)

Sludge oxygen demand (SOD) is estimated by

SOD = 6.52 SL + 9.361

which gives the SOD in mg O_2/hr/g of sludge volatile solids.

Grieves et al. /(1964) also developed kinetic formulae for a plug flow and a completely mixed (CSTR) model. The mathematic mixing models of Cholette and Cloutier (1959) were used in this development. This general treat-

ment follows that of Rich (1963). Additional design data can be obtained
from Eckenfelder (1961).

Steady-state concepts must be applied. The term steady state applied to
activated sludge describes a condition in which the sludge and substrate concentrations exhibit only small variations about average values. Since the
overall reaction is carried out by a mixed population, one may hesitate to
apply the steady-state concept; however, it is quite acceptable. Volterra
(1931) proved a similar assumption to be valid for a population of interpreda-
tory organisms, showing that they will eventually arrive at a quasi/stationary
state in which all organisms remain within fixed limits.

When nutrient concentration C becomes very large, (6) reduces to

$$K = K_{max} \tag{7}$$

Similarly, when C is limiting or much less than C_k, eq. (6) reduces to

$$K = \frac{K_{max}/C}{C_k} \tag{8}$$

Figure 5 is a plot of growth rate versus substrate concentration, yielding the Michaelis-Menten constant C_k and the above equations. If one equates
cellular growth and sludge synthesis, sludge growth can be expressed by:

$$\frac{ds}{dt} = k_1 S \tag{9}$$

In a batch operation at some transition point b, the food concentration C_b is
just sufficient to maintain the maximum multiplication of the bacterial population present (McCabe and Eckenfelder, 1961). Below this point of
discontinuity the growth rate is controlled and is proportional to the limited
concentration of feed. Substrate removal is therefore dependent on substrate remaining and can be expressed by

$$\frac{dc}{dt} = -k_2 C \tag{10}$$

where k_1 = natural logarithmic growth rate

k_2 = natural logarithmic substrate removal rate

Under these conditions there is an increase in organism concentration produced by a unit increase in substrate concentration. At the transition point b the growth rates and removal rates can be expressed in the form

$$k_1 S_b = a k_2 C_b \tag{11}$$

where

$$a = (ds/dt)/(dc/dt) \text{ (Wilson, 1963)}. \tag{12}$$

Assuming (a, k_1) to be constants and

$$k_2 = k'_2 S_b b \tag{13}$$

at point b, then

$$\frac{C_b}{S_b} k_2 = \frac{k_1}{a} \tag{14}$$

or

$$\frac{C_b}{S_b} (k'_2 Sb) = C_b/k'_2 = \frac{k_1}{a} \tag{15}$$

Therefore

$$C_b = \text{constant} \tag{16}$$

C_b should vary as S_b, but experimentally k_2 is independent of C_b (McCabe and Eckenfelder, 1961). If the latter is correct, C should be constant for a given substrate. Reports by Garrett and Sawyer (1952) clearly imply involvement of such constants for specific wastes.

Wilson (1963), in comparing the Monod and Garrett-Sawyer relationships, found K_{max} analogous to k_1 in the two-phase system. Furthermore, as Fig. 5 shows, $C_k = (1/2)C_b$. As C_k is not a function of the organism concentration, additional support is provided for the constancy of C_b.

The effect of initial adsorption on the two phases has been reviewed by McCabe and Eckenfelder (1961) and Wilson (1963). Storage of the transferred substrate has been shown to be inhibited by added chemicals and unacclimatized sludge. These observations provided the involvement of some biological agency. From experimentation, Gaudy and Engelbrecht (1961) have found peak values of metabolic products occurring at the same time as maximum sludge mass. The initial removal of substrate, without a proportionate increase in oxygen uptake, is associated with the production of intermediate products necessary to provide the mass driving force for the various reversible enzyme reactions and for diffusion to new reaction sites. In an oxygen depleted state the sludge formation is low in a high BOD medium. This mechanism provides a satisfactory explanation for the lack of adsorptive power of organisms removed from a high BOD medium and the restoration of this facility by subsequent aeration. The point of discontinuity C_b is therefore not affected by initial removal of substrate. Further work at higher substrate and sludge concentrations have resulted in similar values of C_b. The sludge condition, resulting from the degree of mixing, would have an effect on C_b only if diffusion of substrate to the active microbial centers were limiting. This condition does not arise until the endogenous phase has been reached and flocculation is initiated.

In summary, C_b can be regarded as constant for a given substrate in a suitably mixed reactor. The constancy of C_b is of importance in design work and will be considered later.

C. Complete-Mixing Unit

One of the most significant developments in the field of biological waste treatment has been the complete-mixing modification to the activated sludge process. Early work by Eidsness (1951) demonstrated the significance of intense mixing in activated-sludge plants. Studies with toxic wastes by Hatfield and Strong (1954) prompted the use of continuously fed activated sludge units. The continuously fed units were able to handle higher organic loads than the batch fed systems. The full realization of the advantages of complete-mixing of continuously fed units was not demonstrated until 1958 when McKinney et al. made a study of a full-scale field installation. Full-scale units are more easily described in terms of biochemistry and mathematics of operation (McKinney, 1960, 1962).

Complete mixing is different from the conventional activated sludge system in many ways. In the latter system the untreated wastes are mixed with return sludge at one end of the aeration tank. The microorganisms receive the full impact of any shock load and respond accordingly with sudden increases in oxygen demand and in growth. At the opposite end of the aeration tank the organic load is stabilized and the organism population begins to die off. By definition, the complete-mixing activated sludge process is one in which the untreated wastes are instantaneously mixed throughout the entire aeration tank. As a result of uniform organic loading there is an equally distributed oxygen demand and an equilibrium in biological growth and population.

The chief advantages of the complete-mixing activated sludge unit can be summarized as follows. The process takes a waste of any BOD loading and produces an effluent of any desired BOD concentration in a single stage unit. Other advantages include the ability to absorb shock organic loads, to be unaffected by sludge hydraulic loads, to obtain maximum utilization of the air used, to produce little excess sludge or lots of excess sludge, and to have lower capital costs than conventional activated sludge. Disadvantages include the necessity of periodically emptying the aeration unit to remove nonbiodegradables which have a tendency to build up, and higher power costs for mixing and aerating. Generally, the latter are offset by savings in other areas, and figures indicate a capital cost and operating costs 25-50% below conventional plants. With small treatment plants the economics favor the total oxidation modification of complete-mixing, while larger installations favour high sludge synthesis.

1. Kinetics of Complete-Mixing

There are three basic complete mixing systems: (a) aeration only, (b) excess sludge in effluent, and (c) separate sludge wasting. All units operate on the same principle and are based upon an understanding of the fundamental biochemistry of aerobic biological metabolism. Operation of the units yield an effluent whose soluble BOD depends on the size of the aeration tank and whose suspended solids is a function of the sedimentation tank efficiency.

Bacteria also metabolize organic substrates to produce cellular proto-plasm. The synthesis reactions require energy, which the bacteria obtain by oxidizing a portion of the substrate molecules being metabolized. Although protoplasm is a complex material, it is relatively uniform in its chemical composition. The energy requirements to produce a unit of protoplasm are constant for a given chemical material being metabolized (Burkhead and McKinney, 1968). For each substrate there is a relationship between energy and synthesis.

The relation between energy and synthetic reactions are expressed in the following two equations:

$$\text{organic matter metabolized} \quad = \begin{pmatrix} \text{protoplasm} \\ \text{synthesized} \end{pmatrix} + \begin{pmatrix} \text{energy for} \\ \text{synthesis} \end{pmatrix} \qquad (17)$$

$$\text{net protoplasm accumulation} \quad = \begin{pmatrix} \text{protoplasm} \\ \text{synthesized} \end{pmatrix} - \begin{pmatrix} \text{endogenous} \\ \text{respiration} \end{pmatrix} \qquad (18)$$

Simultaneous solutions to the above equations can be performed if com-mon units are used to describe the terms. An oxygen–equivalent scale is applied wherein concentrations are expressed as five–day BOD. Synthesis and endogenous respiration energies are related to oxygen uptake, and proto-plasm oxygen equivalents are determined from COD analyses.

Expressed in differential form the change in organic matter F_m is equal to the change in the oxygen equivalent of the protoplasm M_s plus the change in oxygen required in synthetic reactions O_s:

$$\frac{dF_m}{dt} = \frac{dM_s}{dt} + \frac{dO_s}{dt} \qquad (19)$$

Likewise the rate of change of protoplasm M_p in oxygen equivalents may be expressed as the change in oxygen equivalent in protoplasm minus the change in oxygen required for endogenous respiration O_e:

$$\frac{dM_p}{dt} = \frac{dM_s}{dt} - \frac{dO_e}{dt} \qquad (20)$$

Equating synthesis and energy terms, the oxygen required for synthetic reac-tion can be expressed as a constant k_1 times the rate of change of protoplasm:

$$\frac{dO_s}{dt} = k_1 \frac{dM_s}{dt} \qquad (21)$$

Eq. (19) can be rewritten as

$$\frac{dF}{dt} = (1 + k_1) \frac{dM_s}{dt} \qquad (22)$$

which expresses the relationship between organic matter metabolized and protoplasm synthesized.

Endogenous respiration has been shown to be a function of the living or active portion of the bacterial mass (McKinney, 1962). Later work by Burkhead and McKinney (1968) has demonstrated endogenous respiration to be primarily a function of sludge age, with varying components of the cell undergoing respiration at different rates. On dying, the cell autolyzes with part being utilized by other microorganisms. A residual fraction remains non-biodegradable, consisting of complex refractile polysaccharide. The endogenous reaction can be assumed as a continuous reaction in which the residual, inert organic matter accumulates in proportion to the endogenous metabolism:

$$\frac{dO_e}{dt} = k_2 M_a \qquad (23)$$

where M_a is the active mass of microbes.

Normally, the declining growth phase is activated by a limiting source of energy. Expressed in oxygen equivalent units, the change in protoplasm is equal to a constant k_4 times the organic concentration F:

$$\frac{dM_s}{dt} = k_4 F \qquad (24)$$

The chief difficulty found with the use of the above equations lies in the evaluation of endogenous respiration, which occurs simultaneously with synthesis. It is not possible to make a direct evaluation of the endogenous respiration reaction in a continuously fed complete-mixing activated sludge system but it is possible to examine endogenous respiration in batch-fed activated sludge systems (Burkhead and McKinney, 1968). The heart of the

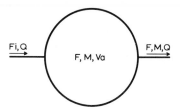

FIG. 6. Aeration only: complete-mixing activated sludge (McKinney et al., 1962).

mathematical approach lies in the evaluation of a series of numerical constants (Smith, 1967). Consideration will be given to the three complete-mixing systems mentioned above.

a. Aeration Only. The simplest form of complete-mixing activated sludge consists of an aeration tank only. Fig. 6 illustrates a unit wherein incoming wastes displace an equivalent volume of mixed liquor. Application of Eqs. (19) through (24) gives the following effluent characteristics and permits a mass balance for aeration only:

$$\underline{input} \quad - \quad \underline{output} \quad - \quad \underline{assimilation} \quad = \quad \underline{accumulation}$$

$$QF_i \quad - \quad QF \quad - \quad k_5 V_a F \quad = \quad V_a \frac{dF}{dt}$$

Dividing by the volume of the aeration tank V_a and substituting for $t = \dfrac{V_a}{Q}$

$$\frac{1}{t} F_i \; - \; \frac{1}{t} F \; - \; k_5 F \; = \; \frac{dF}{dt}$$

At steady state $\dfrac{dF}{dt} = 0$, which permits us to solve for the organic loading F:

$$F = \frac{F_i}{k_5 t + 1} \tag{25}$$

Making a similar balance on active mass and assuming raw wastes to make only a negligible contribution, then

$$\underline{synthesis} \quad - \quad \underset{\text{respiration}}{\underline{endogenous}} \quad - \quad \underline{output} \quad = \quad \underline{accumulation}$$

$$k_6 V_a F \quad - \quad k_7 V_a M_a \quad - \quad Q M_a \quad = \quad V_a \frac{dM_a}{dt}$$

Again dividing by the aeration tank volume and substituting for $t = \dfrac{V_a}{Q}$ and assuming steady state for $\dfrac{dM_a}{dt} = 0$, at equilibrium the mass of

$$M_a = \frac{k_6 F}{1/t + k_7} \tag{26}$$

Volatile solids consist of active mass and inerts generated by the cells in endogenous respiration:

$$M = M_a + M_e$$

A mass balance on endogenous volatile solids reveals

<u>synthesis</u>   −   <u>output</u>   =   <u>accumulation</u>

$$V_a k_8 M_a \quad - \quad Q M_e \quad = \quad \frac{V_a dM_e}{dt} \tag{27}$$

Making steady state assumptions one can solve for M_e. This can be substituted into total volatile solids at equilibrium to give

$$M = M_a \ (1 + k_8 t)$$

If the raw waste feed contains metabolizable volatile solids there usually exists a relationship between volatile solids and organics expressed by

$$M_m = M_{mi}(F/F_i) \tag{28}$$

Similarly, at equilibrium nonmetabolizable inerts can be given by M_i.

At equilibrium, total concentration of volatile suspended solids in the effluent is

$$M = M_a \ (1 + k_8 t) + M_m + M_i \tag{29}$$

Oxygen utilization is related to synthesis and endogenous respiration by

$$\frac{dO}{dt} = k_9 F + k_2 M_a$$

At steady state

$$O = (k_9 F + k_2 M_a)t \tag{30}$$

The oxygen demand of the effluent is simply related to the organic matter discharged and the active mass of microbial solids. When expressed as five-day BOD,

$$\text{Eff. BOD} = F + k_{10}M_a \tag{31}$$

It is easily seen that effluent organic concentration and active mass of bacteria are related to the incoming organic load and the aeration period. Oxygen utilization in the aeration tank is related to the synthesis reactions and to the endogenous respiration.

The aeration-only system is designed to remove between 50 and 75% of incoming BOD with aeration times of from 1 to 4 days. It is not meant to handle high solids loading and has found extensive application in the petroleum industry, especially in the treatment of phenolic wastes.

Problem 1 illustrates many noteworthy characteristics of this operation:

1. The small amount of MLSS has little tendency to settle out in the system.

2. Oxygen demand is low and does not require expensive aeration equipment.

3. The key to the operation is the effluent suspended solids, which are high unless retention time is kept to at least several days.

4. Effluent BOD follows the trend of suspended solids directly.

5. For all practical purposes temperature changes are insignificant to design parameters unless encountered in shock form. It is not uncommon to have settling ponds after the aeration tank, to settle out suspended solids. Equipment design is straightforward, as illustrated in Problem 1.

Problem 1: Aeration - Only Design Illustrations

Analyses on field samples yielded the following data on constants at $20^{\circ}C$ (McKinney, 1962).

k_2 = 0.007 mg/liter of oxygen per liter of active mass per hr.

k_5 = 360 mg/liter ultimate oxygen demand removed per mg/liter ultimate oxygen demand remaining per day.

k_6 = 250 mg/liter volatile active mass per mg/liter ultimate oxygen demand remaining per day, with F as 5-day BOD.

k_7 = 0.144 mg/liter decrease in active mass per mg/liter active mass.

k_8 = 0.036 per day.

k_9 = 7.5 mg/liter oxygen utilization per mg/1.5 day BOD remaining
         per hr.

k_{10} = 0.6 for carbonaceous metabolism or 0.9 for both carbonaceous
         and nitrogenous metabolism.

k_5, k_9 vary with energy-synthesis relationship for different substrates.
         Values used here are for sewage.

The waste to be treated has 250 mg/liter of five-day BOD and is a typical domestic sewage. Assume a one-day retention based on flow and aerator
volume [Eq. (25)].

$$F = \frac{250}{(360)\ (1)\ +\ 1} = 0.69 \text{ mg/liter}$$

Most of the waste is metabolized in 24 hr. Doubling the aeration period
essentially halves the effluent output.

Field units have demonstrated that over a temperature range of 5° to
$35^\circ C$ the rate of biological reactions double with each $10^\circ C$ temperature increase. For all practical purposes F is very small at any reasonable temperature or aeration period [Eq. (26)]:

$$M_a = \frac{250\ (0.69)}{(1/1)\ +\ 0.144} = 150 \text{ mg/liter}$$

Reducing the retention period to 12 hr. raises M_a to 550 mg/liter. Temperature differences of $10^\circ C$ have little effect. With soluble wastes, total
volatile solids accumulation [Eq. (29)] is

$$M = 150\ \left[1 + 0.036\ (1)\right] = 155 \text{ mg/liter}$$

The oxygen demand is

$$\frac{dO}{dt} = 7.5\ (0.69) + 0.007\ (150) = 6.2 \text{ mg/liter } O_2/\text{hr}$$

If the retention time is cut in half, this would raise the oxygen uptake
rate to 14.1 while temperature has little effect. Effluent BOD is given by

$$\text{Eff. BOD} = 0.69 + 0.6\ (150) = 91 \text{ mg/liter carbonaceous five-day}$$
$$\text{BOD}$$

Halving the retention time raises the solids effluent output to 320 mg/ liter while temperature again has little effect.

Hydraulic design is easily calculated by working backward from the desired effluent conditions to give a retention time. For a given flow, this dictates the aeration tank volume.

The mixing and aeration devices depend largely on tank construction and manufactures equipment, with little regard to MLSS as they are quite low.

b. Aeration Plus Sludge Return. The addition of a settling tank and sludge return equipment permits a greater concentration of MLSS in the aeration tank. Figure 7 illustrates the unit, which is now capable of greater metabolic rates. Although similar to extended aeration units in conventional plants, biological differences enhance the units operation.

For purposes of calculating retention time, the aeration tank and sedimentation tank may be treated as a single unit volume V.

A mass balance across a system for organics produces Eq. (25) again:

$$F = \frac{F_i}{k_5 t + 1} \tag{32}$$

A microbial balance shows a slight variance from Eq. (26) with x/t replacing 1/t:

$$M_a = \frac{k_6 F}{(x/t) + k_7} \tag{33}$$

A volatile solids mass balance from endogenous substrates yields

$$M_e = \frac{k_8 M_a t}{x}$$

Since almost complete reduction of metabolizable organics results from the process, the metabolizable volatile suspended solids in aeration tank are

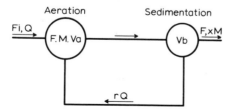

Fig. 7. Aeration plus sludge return -- complete mixing activated sludge (McKinney et al, 1962).

assumed negligible. At equilibrium, inert nonmetabolizable material is given by

$$(M_i)i = xM_i$$

Total volatile suspended solids is therefore

$$M = M_a (1 + \frac{k_8 t}{x}) + \frac{(Mi)\ i}{x} \tag{34}$$

The major problem with the sludge-recirculation system is the value of x, which is related to the maximum MLSS that can be carried in the aeration tank, and to the inert and volatile solids in the influent.

The maximum total MLSS concentration that can be carried in the aeration tank may be calculated from:

$$M_t = \frac{(SDI)\ (10,000)}{\frac{1}{r} + 1} \tag{35}$$

where SDI is the sludge density index and r is the recirculation ratio. Equation (34) can be rearranged to give an expression for x, the fraction of MLSS in effluent:

$$x = \frac{M_a k_8 t + (M_i)\ i}{M - M_a} \tag{36}$$

At equilibrium, the oxygen demand rate is given by

$$\frac{dO}{dt} = k_9 F + k_2\ M_a \tag{37}$$

and the final effluent BOD by

$$Eff.\ BOD = F + x\ k_{10} M_a \tag{38}$$

The simplest operation of this system involves the loss of excess sludge in the effluent. The basic difference in the system is the fact that the effluent suspended solids do not equal the mixed liquor suspended solids, but some value xM. Effluent organic concentration is still dependent on raw wastes concentration as indicated in Eq. (32). The aeration unit carries a maximum MLSS as indicated by Eq. (35).

The above system is designed to remove between 75 and 95% of the incoming BOD with a 24-hr retention time. On startup the effluent will at first appear clear, until buildup of MLSS. At equilibrium the effluent will appear

turbid with up to 150 mg/liter of volatile suspended solids (McKinney, 1962).
It is important to realize that most of these solids are inert and as such are
only a sludge mass handling problem (Kretch, 1963).

Problem 2 illustrates important factors concerning operation and design.
Briefly, (a) the effluent organic concentration is low, and not significantly
affected by retention time or temperature; (b) effluent suspended solids are
high and very much dependent on retention time; (c) active microbial mass is
small in comparison with total volatile solids in the system.

Some difficulties are encountered in the process (Schenk and Howes,
1963). When loading reaches an equilibrium, the active mass entering the
sedimentation unit quickly consumes available dissolved oxygen, leading to
an increase in denitrification. Nitrogen gas released has a tendency to bub-
ble to the surface with small particles of sludge, forming a scum. Rapid
return of settled activated sludge is a good control. Another problem occurs
from the short circuiting of some of the raw wastes. This generally results
in grease and scum in the sedimentation unit which has to be skimmed off.

The requirement for sludge wastage is indicated when (a) the aeration
capacity of the system is exceeded; (b) the return sludge capacity of the system is exceeded, and (c) the final sedimentation unit is upset hydraulically by
high sludge return.

Problem 2: Aeration Plus Sludge Return

Waste is domestic sewage and has 240 mg/liter five-day BOD. For the
flow, assume a retention time of 28 hr, 24 in the aeration tank and 4 in the
sedimentation unit Eq. (32).

$$F = \frac{240}{360 \ (1.16) + 1} = 0.57 \ \text{mg/liter}$$

As in the previous example, F doubles when the aeration period is halved
or if there is $10^{\circ}C$ decrease in temperature. For most extended-aeration
systems, the maximum MLSS is about 6000 mg/liter for an SDI of 1.0
Eq. (35):

$$6000 = \frac{(1.0) \ (10,000)}{(1/r) + 1}$$

Domestic sewage has two-thirds of the total MLSS of volatile solids, or
in this system 4000 mg/liter. Suppose that 50 mg/liter of the volatile suspended solids in the raw waste are inert: then using Eq. (34)

$$4000 = Ma \ 1 + (0.036) \ 1.16/x \ + \ 50/x$$

From Eq. (33) we obtain the microbial mass

$$M_a = \frac{(250) \ (0.57)}{(x/1.16 + 0.144)}$$

Simultaneous solution of the above two equations gives $x = 0.026$ and $M_a = 820$ mg/liter. It is evident that active mass is only 20% of the total volatile MLSS in the system.

The rate of oxygen demand is

$$dO/dt = 7.5 \ (0.57) + 0.007 \ (820) \ = \ 10.015 \ mg/liter$$

Similarly, the effluent BOD is

$$0.57 + 0.026 \ (0.6) \ (820) \ = \ 13.4 \ mg/liter \tag{39}$$

Volatile, suspended solids are

$$xM \ = \ 0.026 \ (4000) \ = \ 104 \ mg/liter.$$

The high solids and low BOD are typical of effluents from this type of plant.

Design work is straightforward, but two things must be emphasized. For a good system, sedimentation efficiency must be high, and mixing devices should be adequate to work in heavy sludges in the aeration tank. Aeration equipment must be free from clogging.

c. Separate Sludge Wastage. To produce an effluent with the least suspended solids and to continually remove some of the refractile organic particulates, separate sludge wasting is necessary. Figure 8 illustrates the system wherein only part of the flow Q goes out the effluent. The wasted and effluent suspended solids are given by SM and xM, respectively.

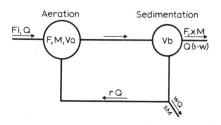

Fig. 8. Separate sludge wasting -- complete mixing activated sludge (McKinney et al., 1962).

Assuming the sludge wasted to have a negligible organic concentration, a mass balance on substrate yields equations similar to (25) and (32).

A general mass balance across the system gives

accumulation = synthesis - endogenous - outflow in - outflow in
                           respiration   effluent   waste sludge

$$\frac{V_d M_a}{dt} = V k_6 F - V k_7 M_a - Q (1-w) \times Ma - Q(w)SM_a$$

where Q is the rate at which organic matter is added to the aeration tank and W is the fraction of Q wasted. Dividing by Q, and assuming that $t = v/Q \; dM_a/dt = 0$ at steady state, then

$$M_a = \frac{k_6 F}{k_7 + (1/t) (x-wx + wS)} \tag{40}$$

The value of S is related to the settling characteristics of the sludge and is given by Eq. (41) (McKinney, 1962):

$$S = \frac{(SDI) \; 1000}{mt} \tag{41}$$

A mass balance for volatile suspended solids gives:

accumulation = synthesis by   - outflow in - outflow in
               endogenous respiration   effluent   sludge waste

$$V (dM_e/dt) = V k_8 M_a - Q (1-w) \times Me - QwSMe$$

This can be solved to yield

$$M_e = \frac{k_8 M_a t}{x + Sw - wx}$$

Analyses and expression of inert volatile solids in feed gives

$$(Mi) \; i = x M_i (1-w) + SM_i w$$

or

$$M_i = \frac{(Mi) \; i}{x + Sw - wx}$$

Total volatile suspended solids is therefore given by

$$M = M_a + M_e$$

or

$$M = M_a \left(\frac{1 + k_8 t}{x + Sw - wx}\right) + \frac{(Mi)\, i}{x + Sw - wx} \tag{42}$$

Oxygen demand rate is

$$dO/dt = k_9 F + k_2 M_a \tag{43}$$

where the effluent BOD is the same as Eq. (39).

It is readily seen that the choice of x and w produces an infinite number of solutions to these equations. The separate sludge wasting system is similar to a conventional plant, with chief advantages in uniform oxygen demand and biological consistency. Units of this type can achieve over 99% BOD reduction with low suspended solids in the effluent.

Problem 3: Separate Sludge Wastage

As in Problem 2, F = 0.57 mg/liter. For a typical operation, Mt = 2000 mg/liter and SDI = 1.0, from which

$$S = 1 \times 10000/2000 = 5$$

x varies between 0.005 and 0.01 depending on sedimentation tank efficiency. Selecting a value for x = 0.005, and with t = 1.16 days, use Eq. (37) to obtain the active cell mass:

$$M_a = \frac{250\,(0.57)}{(1/1.16)\,(0.005 - 0.005\,x + 5\,w) + 0.144}$$

If the MLSS is two-thirds volatile, M is 2/3 (2000) or 1330 mg/liter.

$$1330 = M_a \left(\frac{1 + 0.036\,(1.16)}{0.005 + 5w - 0.005w}\right) + \frac{50}{0.005 + 5w - 0.005w}$$

Simultaneous solution of the above two equations yields w = 0.019 and M_a = 620 mg/liter.

Reducing the retention period by one-half causes w to increase to 0.023 and reducing by three-quarters increases w to 0.029. Waste sludge does not increase rapidly with a decrease in retention period. Greater reduction in w can be obtained by using a primary settler.

Larger sludge wastage is occasioned by lowering the MLSS in the aeration tank, lowering the temperature, or increasing the waste loading. The oxygen demand rate is

$$dO/dt = 7.5 \ (0.57) + 0.007 \ (620) = 8.6 \text{ mg/liter hr}$$

and the effluent BOD is

$$\text{Eff. BOD} = 0.57 + 0.005 \ (0.6) \ (620) = 2.43 \text{ mg/liter}$$

Effluent suspended solids is simply xM volatile times $\frac{3}{2}$:

$$0.005 \ (1330) \cdot \frac{3}{2} = 10 \text{ mg/liter}$$

Problem 3 relates sludge wastage to retention time, which is not always significant. The sedimentation tank is again the limiting device, with the same restrictions mentioned for the aeration plus sludge return unit.

As previously discussed, the heart of the mathematical approach lies in the evaluation of a series of numerical constants k_5 through k_{10}. Recent work by Burkhead et al. (1968) has established new energy synthesis constants k_1. The values found have been specific to the substrate employed. Results also demonstrate distinct protoplasm compositions for specified carbohydrate wastes and noncarbohydrate wastes. Table 2 relates these compositions to oxygen demands.

Constants k_5 and k_6 are dependent on the value of k_1 and its interrelationships. Figures 9(a) and 9(b) illustrate this dependence in relation to the protoplasm of the biological mass.

It is important to emphasize that k_1 can be evaluated for a complex waste composition by

$$k_1 \text{ complex} = \Sigma \ xi. \ ki$$

where

xi is the % COD of the waste

ki is the energy-synthesis constant for component i

The only limitations to the use of this procedure are in the analytical determinations on the raw wastes since their biodegradable portions must be known. A procedure by Symons et al. (1960) is recommended.

TABLE 2

OXYGEN EQUIVALENCE VALUES OF VARIOUS PROTOPLASM COMPOSITIONS

McKINNEY ET AL. (1968)

Protoplasm formulation		Molecular weight	O_2 Equivalence of protoplasm (mg/liter ultimate O_2 per mg/liter protoplasm)
$C_5H_7O_2N$	Noncarbohydrate	113	1.42
$C_5H_9O_3N$	Noncarbohydrate	131	1.22
$C_5H_9O_3N$	Carbohydrate	131	1.22
$C_5H_{11}O_4N$	Carbohydrate	149	1.07

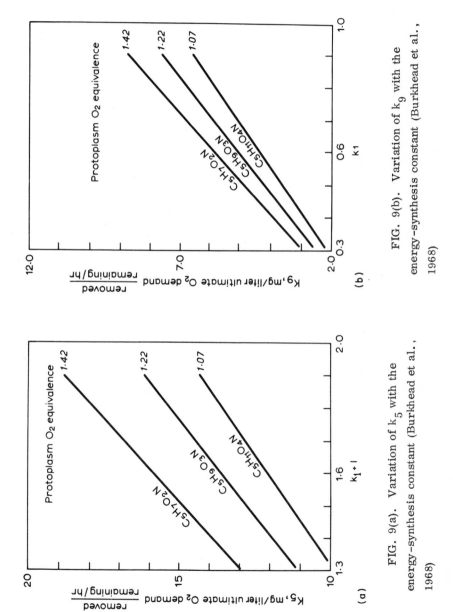

FIG. 9(b). Variation of k_9 with the energy-synthesis constant (Burkhead et al., 1968)

FIG. 9(a). Variation of k_5 with the energy-synthesis constant (Burkhead et al., 1968)

Table 3 illustrates the advantages, disadvantages, and application of all three types of complete-mixing activated sludge units.

IV. REACTOR DESIGN THEORY

In the bio-oxidation of organic wastes, enzymatically catalyzed chemical reactions occur wherein substrates are transformed to a number of products. An understanding of the biochemical reaction kinetics occurring and elementary reactor design are essential to the design engineer if he is to achieve the most economical processing of wastes.

From reaction kinetics, the order of a reaction is by definition the sum of the powers to which the concentrations of the reactants are raised (Levenspiel, 1962).

Three of the more common ideal reactors are shown in Fig. 10. The batch reactor is an unsteady state unit, with composition changing continuously with time. The plugflow and backmix flow reactors are steady-state units, with constant feed and product compositions. In principle the performance of any reactor can be predicted if the reaction kinetics and reactor geometry are known. The general approach to all design work begins with a material balance over the reactor.

Comparison of Eqs. (45) and (46) reveals that the only difference between the plugflow and the backmix reactors is that $-T_A$ is constant in the latter but varies throughout the plugflow reactor. If the mass density of the fluid remains constant (t = λ), then Eq. (45) equals Eq. (46).

In batch systems, holding time is used exclusively as the capacity measure of a reactor. For flow systems, space time is more convenient, and in cases where fluid density changes it gives the reactor volume directly, where holding time does not.

Consider the elementary reaction

A $\longrightarrow$ Products

If a plot of C_A versus time gives a straight line, the reaction is said to be of

TABLE 3

COMPARATIVE STUDY OF THE THREE TYPES OF COMPLETE-MIXING

Item \ Unit	Aeration only	Aeration plus sludge return	Separate sludge wasting
Effluent BOD,	50-75% Reduction	75-95% Reduction	99% Reduction
Half-retention t,	Large effect	Little effect	Effect
Change T – 10°C	Little effect	Little effect	Little effect
Active mass	High	Very low	Very low
Half-retention t	Large effect	Effect	Effect
Change T – 10°C	Little effect	Little effect	No effect
O_2 demand	Very low	High	High
Half-retention t	Effect	Large effect	Large effect
Change T – 10°C	Little effect	Little effect	Little effect
Inert Volatile solids	High	High	Low
Half-retention t	Large effect	Large effect	Effect
Change T – 10°C	Little effect	Effect	Effect

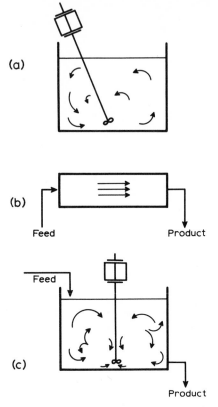

FIG. 10. Three types of ideal reactors (a) Batch (b) Plug Flow (c) Back-mix (Levenspiel, 1962).

zero order, and its rate can be given by

$$-r_A = \frac{dC_A}{dt} = k C_A^n$$

where $n = 0$. The rate of conversion of A is therefore independent of the concentration of materials.

If a plot of log C_A versus time gives a straight line, the reaction is said to be of first order, and its rate can be given by

$$-r_A = \frac{-d C_A}{dt} = k C_A^n$$

where $n = 1$. As shown above, n is the order of the reaction and can be of any magnitude.

Reactor Design Equations

General material balance is based on an elemental volume dV:

$$\underset{\text{in}}{\underline{\text{Reactant flow}}} = \underset{\text{out}}{\underline{\text{Reactant flow}}} + \underset{\text{reaction}}{\underline{\text{Reactant loss due to}}} + \underset{\text{accumulation}}{\underline{\text{Reactant}}}$$

For a uniform concentration within the reactor, dV can be enlarged to include the entire reactor.

1. Batch reactor

The first two terms of the material balance are zero. Thus, at any instant the molal reaction rate is

$$-r_A = \frac{dC_A}{dt} = C_{A0} \frac{dX_A}{dt}$$

In terms of residence time t for a batch run,

$$t = C_{A0} \int_0^{X_{AF}} \frac{dX_A}{-r_A} \tag{44}$$

2. Plugflow reactor

For a plugflow reactor, residence time t is generally replaced by space time τ, to allow for varying mass density throughout the reactor. If the fluid density remains constant, τ can be equated to t.

$$t = \frac{V}{v} = \frac{V C_{A0}}{F_{A0}} = \tau$$

In the plugflow reactor, steady state exists and accumulation can be set equal to zero for the material balance equation for an elemental volume dV:

$$F_A = (F_A + dF_A) - r_A dV$$

However,

$$dF_A = d\left[F_{A0} (1 - X_A)\right] = F_{A0} \, dX_A$$

or

$$F_A = F_A - F_{A0} \, dX_A - r_A dV$$

and

$$\frac{dV}{F_{A0}} = \frac{dX_A}{-r_A}$$

Integrating:

$$V = F_{A0} \int_0^{X_{Af}} \frac{dX_A}{-r_A}$$

or

$$\tau = C_{A0} \int_0^{X_{Af}} \frac{dX_A}{-r_A} \tag{45}$$

3. Backmix flow reactor

In a backmix unit, composition is uniform and an overall material balance can be made.

$$F_{A0} = (F_{A0} - dF_A) - r_A V$$

$$F_{A0} = F_{A0} (1 - X_A) - r_A V$$

and

$$V = \frac{F_{A0} X_A}{-r_A}$$

or

$$\tau = \frac{C_{A0} X_A}{-r_A} \tag{46}$$

The significance of space time and the kinetic order of a reaction become apparent in choosing a reactor for a specific application. Reactor design controls two factors that may profoundly influence the economics of the overall process. These two factors are the reactor size and the distribution of products of reaction. By reactor size is meant the volume required for the reactive materials. Product distribution is a prime consideration in multiple reaction systems. A reactor with the appropriate type of flow will maximize the production of the desired product and at the same time depress formation of undesired materials.

Before comparing flow reactor systems it should be made clear that a

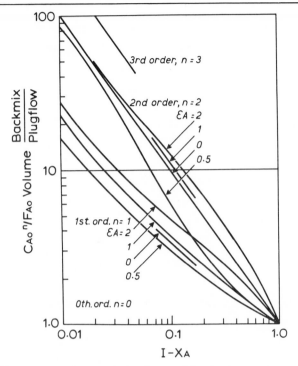

FIG. 11. Performances of plugflow and backmix reactors (Levenspiel, 1962).

batch reactor entails more than residence time. Shutdown times for cleaning and refilling should be added to retention time. Unless small amounts of many different products are desired, the flexibility of the batch unit gives way to the continuous plugflow reactor.

If an assumption is made as to the regularity of the waste stream, subsequent comparisons will be restricted to the backmix and plugflow reactors.

In general, the ratio of sizes of backmix and plugflow reactors will be influenced by the extent of reaction, the reaction type, and the rate of equation. Figure 11 illustrates the ratio of volumes for the backmix and plugflow reactors for identical feed composition, C_{A0}, and flow rate, F_{A0}, at any specified conversion, X_A. The figure shows the following:

1. For zero-order reactions, reactor size is independent of flow. In all others the backmix reactor required is greater than the plugflow, the ratio increasing with the order of the reaction.

2. Greater conversion further increases the volume ratio.

3. E_A refers to the fractional change in volume of reactant A on going to product. It is seen that a density decrease or expansion during the reaction increases the volume ratio.

It is evident from the above that knowledge of the reaction occurring is most valuable in the selection of waste treatment equipment.

V. CONTACT STABILIZATION

Sewage or industrial waste with a very high BOD level requires adoption of the contact stabilization process. This process converts the standard aeration basin (contact time 4-8 hr) into a biosorptive chamber (contact time 0.5-1.0 hr) where organic insols and organic solubles are adsorbed and absorbed and passed to a settler. The solids are then pumped to a contact stabilizer and aerated to remove the organic captured in the sorbtive process. Hold up time may be 3-6 hr. Excellent control must be maintained in the sludge aeration basin as the sludge must be conditioned to just the right point (Borrough, 1968). A contact stabilization process operates with about three times the concentrations of mixed liquor solids as a convention process. The concentration of intracellular substrates at equilibrium fits a Langmuir adsorption isotherm as follows:

$$G_{in}^{eg} = Y \frac{G_{ex}}{G_{ex} + K_m} \tag{47}$$

Where G_{in}^{eg} is the intracellular substrate

G_{ex} is the initial extracellular substrate concentration

K_m is the apparent dissociation constant

Y is the specific capacity, i.e. the maximum amount of substrate the cell (sludge) can accumulate.

The kinetics for the initial uptake of organic is expressed by

$$G_{in} = \frac{V_{in}^{max}}{k_{ex}} \cdot \frac{G_{ex}}{G_{ex} + K_m} \quad 1 - e^{-k_{ex}t} \tag{48}$$

Where G_{in} is the concentration of substrate accumulated at any time t

V_{in}^{max} is the maximum velocity of uptake

k_{ex} is the exit rate constant

The steady state condition can be described by

$$G^{eg} = \frac{V_{in}^{max}}{k_{ex}} \cdot \frac{G_{ex}}{G_{ex} + K_m} \tag{49}$$

The equilibrium substrate concentration in the substrate pool inside the cell is described by Eq. (44). This equation also describes the kinetics of transport of substrate (waste) to this pool which predicts an increase with increasing substrate concentration. This increases until transport enzymes become saturated. Above the saturation level only the equilibrium level of substrate concentration in the inside pool increases.

NOMENCLATURE

C_A	molar concentration of reactant A
C_{A0}	initial molar concentration of reactant
ΔE	activation energy
E_{ff} BOD	effluent 5-day BOD
F	organic concentration
F_A	molar flow rate of substance A
F_{A0}	initial molar flow rate of substance A
F_i	organic matter in the raw wastes
F_m	ultimate oxygen demand of the organic matter metabolized
K	equilibrium constant for a reaction
k_1	energy synthesis constant
k_2	endogenous respiration constant
k_3	bacterial growth rate in an unlimited supply of organic matter
k_4	growth rate when food becomes limiting
k_5	assimilative oxygen equivalence constant
k_6	synthesis oxygen equivalence constant
k_7	endogenous respiration oxygen equivalence constant
k_8	volatile solids synthesis constant
k_9	oxygen utilization synthesis constant
k_{10}	metabolism constant over a five-day period
M_p	net protoplasm accumulation as oxygen equivalent
M	total mass of volatile suspended solids
M_a	active or living mass of micro-organisms
M_e	volatile suspended solids formed from endogenous metabolism, and endogenous oxygen uptake
M_i	ion concentration of inert solids
M_m	matabolized volatile suspended solids in aeration tank and effluent
M_{mi}	metabolizable volatile suspended solids in the raw wastes
M_s	oxygen equivalent of the protoplasm synthesized
M_t	total MLSS concentration
n	order of the reaction
O	total oxygen utilized
O_e	endogenous oxygen uptake
O_s	oxygen uptake for synthesis

Q	rate at which organic matter in raw wastes is added to aeration tank
R	gas constant
r	recirculation ratio
r_A	molal reaction rate for component A
S	fraction of MLSS in sludge wasted
SDI	sludge density index
T	temperature
t	time
v	volumetric flow rate
V_a	volume of aeration tank
V	total volume of system
w	fraction of flow Q that is wasted
x	fraction of MLSS in effluent
X_A	fractional conversion of component A
X_{Af}	final fractional conversion of component A
E_A	Fractional change in volume of component A
λ	space time

REFERENCES

Aiba, S., A. E. Humphrey, and N. F. Millis. 1965. Biochemical Engineering. Academic, New York.

Borrough, P. C. 1968. "Contact Stabilization." Process Biochemistry, Jan. 35-37.

Burkhead, C. E. and R. E. McKinney. 1968. "Application of complete-mixing activated sludge design equations to industrial wastes." J. Water Pollution Contr. Fed. 40, 4457.

Cairns, J. Jr. 1956. "Effects of increased temperature on aquatic organisms." Proc. 10th Ind. Waste Conf. Purdue University. Ext. Ser. 89, 346.

Chollette, A. and L. Cloutier. 1959. "Mixing efficiency determinations for continuous flow systems." Can. J. Chem. Eng. 37, 105-112.

Cooke, W. B. and F. J. Ludzack. 1958. "Predacious fungus behaviour in activated sludge systems." Ser. Ind. Wastes 30, 1490.

Eckenfelder, W. W. Jr. 1961. "Designing biological oxidation systems for industrial wastes--I, II, and III." Waste Eng. May 238-257, June 292-294, July 344-347.

Eidsness, F. A. 1951. "III. The aero accelerator pilot plant studies." Sew. and Ind. 23, 7, 843.

Ford, D. L. and W. W. Eckenfelder Jr. 1967. "Effect of process variables on sludge floc formation and settling characteristics." J. Water Pollution Contr. Fed. 39, 11, 1850.

Garrett, T. M. and C. N. Sawyer. 1952. "Kinetics of removal of soluble BOD by activated sludge." Proc. 7th Ind. Waste Conf., Purdue University p. 51.

Gaudy, A. F. and R. S. Englebrecht. 1961. "Quantitative and qualitative shock loading of activated sludge systems." J. Water Pollution Contr. Fed. 33, 800.

Gram, A. L. III. 1956. "Reaction kinetics of aerobic biological processes." Thesis, Dept. of Engineering, Univ. California, Report No. 2, I.E.R. Series 90.

Greeley, S. A. 1945. "The development of the activated sludge method of sewage treatment." Sew. Wastes J. 17, 1137.

Grieves, R. B., W. F. Milbury and W. O. Pipes, Jr. 1964. "A mixing model for activated sludge." J. Water Pollution Contr. Fed. 36, 619-635.

Hoover, S. R. and N. Porges. 1952. "Assimilation of dairy wastes by activated sludge. II. The equation of synthesis and rate of oxygen utilization." Sew. and Ind. Wastes 24, 306.

Ingraham, J. L. and J. L. Stokes. 1959. "Psychrophilic bacteria." Bacteriol. Rev. 23, 97.

Hatfield, R. and E. Strong. 1954. "Small-scale laboratory units for continuously fed biological treatment experiments. I. Aeration units for activated sludge." Sew. and Ind. Wastes 26, 10, 731.

Kretch, C. H. 1963. "Operation of an underloaded activated sludge plant." Senior Sewage Works Operator Manual, Ontario Water Res. Comm., Toronto, Ontario.

Levenspiel, O. 1962. Chemical Reaction Engineering. Wiley, New York.

Ludzack, R. J. R. B. Schaffer, and M. B. Ettinger, 1961. "Temperature and feed as variables in activated sludge performance." J. Water Pollution Contr. Fed. 33, 141.

McCabe, B. J. and W. W. Eckenfelder, Jr. 1961. "B.O.D. removal and sludge growth in the activated sludge process." J. Water Pollution Contr. Fed. 33, 3,258.

McKinney, R. E. 1960. "Complete mixing activated sludge." Water and Sewage Works 107, 69.

McKinney, R. E. 1962. "Mathematics of complete-mixing activated sludge." J. San. Eng. Div., Proc. Am. Soc. Civ. Eng. 88, 87.

McKinney, R. E. and A. Gram. 1956. "Protozoa and activated sludge." Sew. and Ind. Wastes 28, 1219.

McKinney, R. E., J. M. Symons, W. G. Shifrin and N. Vezina. 1958. "Design and operation of a complete mixing activated sludge system." Sew. and Ind. Wastes 30, 3287.

Monod, J. 1949. "The growth of bacteria cultures." Ann. Rev. Microbiol, 3, 371.

Norman, J. D. and A. Busch. 1966. "Application of research information to design and operation of industrial waste facilities." 13th Ontario Ind. Waste Conf. Ontario Wat. Res. Comm. p. 157.

Popel, F. 1962. "Determination of the size of tanks for the aerobic and anaerobic degradation of organic wastes." Proc. 1st Int. Conf. Adv. Water Poll. Res., (London) 2, 451-481.

Porges, N., L. Jasewicz, and S. R. Hoover. 1955. "Biochemical oxidation of dairy waste, VII: Purification, oxidation, and storage." 10th Ind. Waste Conf. Purdue Univ., 89, 135.

Rao, B. S., and A. F. Gaudy, Jr. 1966. "Effect of sludge concentration on various aspects of biological activity in activated sludge." J. Water Pollution Contr. Fed. 38, 5, 794.

Rich, L. G., 1963. Unit Processes of Sanitory Engineering. Wiley, New York.

Rose, A. H. 1968. "Physiology of microorganisms at low temperatures." J. Appl. Bacteriol. 31, 1.

Schenk, C. F. and C. J. Howes, 1963. "Significance and identification of protozoa in activated sludge." Senior Sewage Works Operators Manual, Ontario Water Resources Commission, Toronto, Ontario.

Smith, H. S. 1967. "Homogeneous activated sludge 2; principles and features of the homogeneous system." Water and Wastes Eng. Aug., 56-59.

Stanier, R. Y., M. Dorudoroff, and E. A. Adelberg. 1963. The Microbial World. Prentice-Hall, 2nd Ed., Englewood Cliffs, New Jersey.

Stokes, J. L. and J. M. Larkin. 1968. "Comparative effect of temperature on the oxidative metabolism of whole and disrupted cells of a psychrophilic and mesophilic species of Baccillus." J. Bacteriol, 95, 95.

Symons, J. M., R. C. McKinney, and H. H. Hassis. 1960. "Industrial Wastes - a procedure for determination of the biological treatability of industrial wastes." J. Water Pollution Contr. Fed. 32, 8, 841.

Theriault, E. J. and P. D. McNamee. 1936. "Adsorption by activated sludge." Ind. Eng. Chem. 28, 79-82.

Volterra, V. 1931. Lecons sur la Theorie Mathematique de la Lutte pour la Vie. Gauthier-Villars et Cie, Paris, France.

Wilson, I. S. 1963. "Concentration effects in the biological oxidation of trade wastes." Int. J. Air Water Pollution 7, 99.

TRICKLING FILTERS

Trickling filters are used to process large volumes of sewage or organic industrial waste. They are not filters in the classical sense since they do not remove insoluble solids, but they do remove soluble organics by bio-oxidation procedures. Trickling filters consist in shallow or deep tanks packed with porous media, in which the influent waste is sprayed or trickled over the surface. Operating depth is critical and most units will have a minimum depth of 4 ft. Standard units are constructed with about 1 ft of depth per 6 ft of diameter. A complex biological slime forms on the surface of the packing media, and the organic waste coming in contact with the slime is oxidized to CO_2 and H_2O. Some of the energy produced in the oxidation is utilized in producing additional cellular materials and new slime. Biological slime is continuously sloughed off and it is usually collected in a secondary clarifier. Inorganic materials are also removed by the microbes but this is not normally a primary function of a trickling filter.

The slime layer is coated on the porous media as shown in Fig. 1. The process is aerobic, with the oxygen in the process supplied by the dissolved oxygen present in the waste and that transferred from the air present in the voids to the liquid film. A thin anaerobic layer builds up as the slime coat thickens, which aids in the sloughing off of old slime by scouring caused by

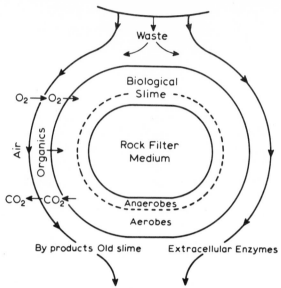

FIG. 1. Bioslime formation on packing media.

influent flow. Organics are easily adsorbed and oxidized and some of them may
be even oxidized in the influent itself through the agency of specific extracellular enzymes, which are exccreted from the slime mass. A general list
of microbes and flora existing in slime is presented in Table 1.

TABLE 1

MICROBES PRESENT IN SLIME FROM TRICKLING FILTERS[a]

Fungi and Bacteria

Mycelia sterilia
Actinomyces flavus
Actinomyces rectus
Fusarium sp.
Pseudomonas sp.
Achromobacter sp.
Zoogloea ramigera
Beggiatoa alba
Flagellata
Amoebae
Ciliata   I.   Free forms
                                 Amphileptus claparedei
                                 Aspidisca costa

Ciliata cont'd.

		Aspidisca lynceus
		Chilodonella uncinata
		Colpidium colpoda
		Euplotes charon
		Glaucoma scintillans
		Halteria cirrifera
		Litonotus fasciola
		Oxytricha fallax
		Paramecium caudatum
		Paramecium putrinum
		Tellina magna
		Metopus es
		Uronema marinum
		Lacrymaria sp.
	II.	Fixed forms
		Opercularia coaritata
		Vorticella microstoma
		Vorticella sp.
Suctoria		Podophrya fixa
		Tokophrya sp.
Nematoda		
Rotatoria		Cephalodella sp.
		Habrotrocha bidens
		Habrotrocha trispus
		Lecane clara
Oligochaeta		Aeolosoma sp.
		Marionina argentea
Copepoda		
Acarina		
Insecta		Collembola
		Diptera (Psychoda sp.)
Algae		Cyanophycae
		Chlorophycae
		Bacillariophyta

[a] After Munteanu et al, 1967.

A typical trickling filter consists of a primary clarifier (Fig. 2) to remove insoluble solids that settle rapidly. Nonremoval will clog the trickling filter. The trickling filter follows with the effluent from this unit passing to a secondary clarifier. Recirculation may be directly from the trickling or from the secondary clarifier back into the influent feed to the trickling filter. A cutaway of a trickling filter is shown in Fig. 3. Highest rates of BOD removal on trickling filters usually occur at 34-38 °C.

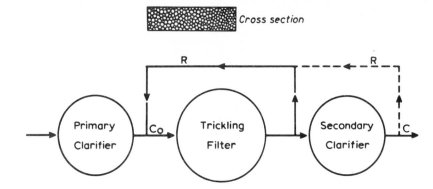

FIG. 2. Single-stage trickling filter.

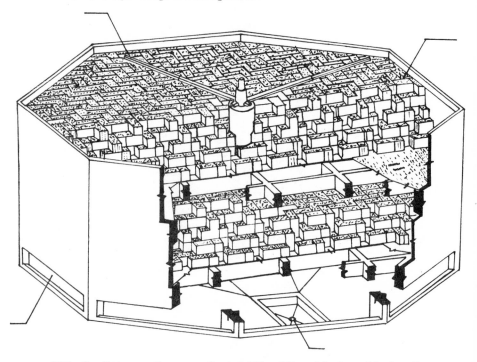

FIG. 3. Cutaway diagram of a trickling filter (block packing), after Minch et al., 1962.

Normally, pH of the influent is not a critical factor in trickling filter operation. The slime will adapt to a pH range ranging from 4.5 to 9.5; however, pH conditions oscillating from low to the high and back will have a detrimental effect on bio-oxidation.

One of the most important characteristics of a trickling filter is the ability to withstand shock-loads of organic and toxic wastes. This may be

attributed to the short residence time the waste has in the filter. Continued
exposure of the slime to the toxicant will result in complete malfunction or
possibly adaptation by the microbial slime to the toxicant.

Ideal packing material should:

1. Offer a surface on which the biological slime can grow successfully.
2. Permit the waste liquid to flow over the bios complex in as thin a film as
   possible to allow for rapid oxygen transfer from the air.
3. Have adequate void space so that a steady supply of air reaches the film.
4. Have adequate void space to allow free passage of treated effluent and free
   removal of bios film scoured from the packing.
5. Be inert and chemically stable.
6. Take loadings up to 1 lb BOD/yd^3 of packing.

I. FILTER MEDIA

The packing medium has changed rapidly in recent years. Crushed rock,
clinker, blast furnace slag, and asbestos are common, but a host of other materials have been used, e.g., vitrified tile, raschig rings (ceramic), Berl saddles,
Dowpac (10 & 20), glass spheres, Mead-Cor, polygrid, and porcelain spheres
(Table 2). Plastic media generally possess greater surface-area-to-volume
ratios and permit better aeration due to increased porosity (90% void space)
and to a more uniform shape.

Filters are typically of concrete or steel construction. In terms of
functionality, the packing is most important, although hydraulic loading and
BOD loading are critical in filter design and sizing.

Most pilot models use spherical packing. In practice, media are treated
as approximating cubics ranging in size from 2-3 1/2 in. Underdrains of rock
filters are perforated tile blocks through which effluent passes and air circulates. Filters are encased with circular concrete walls and rotary distri-
butors are used for discharging waste to surface. Feed is continuous or intermittent. Trickling filters are designated as high rate or standard. High rate
filters take a hydraulic load of $9-45 \times 10^6$ gal/day/acre with a biological load in
the range of 25-300 lb/day/1000 ft^3. Recirculation is quite common. Standard rate trickling filters are operated at 1-5 mgd/acre with organic loads
of 5-25 lb $BOD/day/1000$ ft^3. Recirculation is usually not used on standard
rate trickling filters.

TABLE 2

HYDRAULIC CHARACTERISTICS OF SELECTED FILTER MEDIA[a]

Media	A_v[b] (ft^2/ft^3)	Exponent n	Coefficient C
Polygrid	30	0.65	9.5
Glass spheres, 0.5 in. (2.8 cm) diam	85	0.82	22.5
Glass spheres, 0.75 in. (1.9 cm) diam	60.3	0.80	15.8
Glass spheres, 1.0 in. (2.5 cm) diam	41.6	0.75	12.0
Porcelain spheres, 3.0 in. (7.6 cm) diam	12.6	0.53	5.1
Rock, 2.5 to 4.0 in. (6.3 -- 10.2 cm)[c]	---	0.408	4.15
Dowpac	25	0.50	4.84
Asbestos[d]	25	0.50	5.10
Mead-Cor[e]	30	0.70	5.6
Asbestos[f]	50	0.75	7.2
Asbestos[g]	85	0.80	8.0

a – After Eckenfelder and Barnhard, 1963.

b – Specific surface. Multiply by 3.32 to obtain m^2/m^3 (Area/Volume).

c – Reported as model time.

d – Corrugated asbestos packing

e – The Mead Corp.

f – B. F. Goodrich Co.

g – Johns-Manville Corp.

Most filters operate a few degrees below the ambient air temperature. The predominant microflora in the slime will be mesophilic, possibly psychrophilic, or even thermophilic depending on climatic conditions and the rate of air circulation. At a flow rate of 6 gal/mm/ft^3, an influent temperature of 48 °C is reduced to 38 °C at a depth of 10 ft. When the feed rate is cut in half, only a 7 ft depth is required to reduce the temperature to $38°C$.

II. TRICKLING FILTER DESIGN

Three basic empirical design approaches are being used for trickling filters (Baker and Graves, 1968). These are (a) National Research Council (NRC) formula (1946), (b) the Eckenfelder formula (1963), and (c) the Galler and Gotaas formula (1964). The design engineer should be acquainted with the strengths and weaknesses of each of these approaches and be aware that more formal mechanistic models are forthcoming. Also, an effort should be made to examine the original papers cited herein to obtain figures and nomographs for making estimates.

The first good trickling filter formula and one still widely used is the NRC formula. It is based upon data and evaluations made at units installed at military installations.

$$E = \frac{1}{1 + C \left(\dfrac{W}{VF} \right)^{0.5}} \tag{1}$$

where
    E is the fraction of BOD removed
    W is the influent organic load (lb/day)
    V is the filter vol (1000 ft^3)
    F is the recirculation factor
    C is the constant (0.0561)

The organic loading W is actually a product of the flow rate Q, the influent BOD (L_i expressed in milligrams per liter), and a constant 8.34.

the recirculation factor F is defined as

$$F = \frac{1 + R}{1 + (1 - p)R^2} \qquad (2)$$

where   R is the recirculation ratio

p is the weighting factor (≈ 0.9)

The recirculation ratio is the ratio of the amount of recycle flow to plant influent flow. Where second stage filters are used there is a noticeable "decrease in treatability" in the effluent waste from the first filter. The magnitude of this factor is dependent upon the amount of removal of BOD in the first stage. This retardant effect is handled by introducing a factor to retard organic loading equal to $\dfrac{1}{1 - E_1^2}$ in which E_1 is the efficiency of the first stage. Thus, the equation for the second stage is

$$E_2 = \frac{1}{1 + 0.0561 \left(\dfrac{W_2}{VF(1 - E_1)^2} \right)^{0.5}} \qquad (3)$$

where E_2 is the efficiency of second stage

W_2 is the organic loading of influent to second stage

These equations are based upon the efficiency of the trickling filter and the settling tanks following the trickling filter. The equation for the second stage filter is based on the existence of a settler between first and second stage filters.

Velz (1948) proposed a design formula for BOD removal based on the principle that the rate of extraction of organic matter per interval of filter depth is proportional to the remaining concentration of organic matter as measured in terms of its removeability.

$$\frac{L_D}{L} = 10^{-KD} \qquad (4)$$

where   L is the total removable fraction of BOD

L_D is the corresponding quantity of removable BOD at depth D

Schulze (1960 a, b) modified this equation somewhat to account for hydraulic loading and a temperature correction factor and for recirculation. Prior to this, Howland (1953, 1958) became concerned with the expression of the time required for a fluid to travel through a trickling filter bed and expressed this time X as

$$X = \frac{(1.035^{T-20}) D}{(Q/A)^{2/3}} (1 + R)^{1/3} \tag{5}$$

where   X is the time of travel

T is the temperature (C^o)

A is the surface area of bed

In 1960 Schulze developed a similar expression:

$$\frac{Le}{Li} = 10^{-\frac{KD}{(Q/A)^{2/3}}} \tag{6}$$

where   Le is the BOD in the filter effluent (mg/liter)

Li is the initial BOD (mg/liter)

D is the filter depth (ft)

Q is the hydraulic load (10^6 gal/acre/day)

K is the constant

The temperature correction factor was

$$\frac{L}{Li} = 10^{-b} K_{20} \left(\frac{D}{Q^{2/3}} \right)$$

$$b = 1.035^{T-20}$$

The recirculation of the combined hydraulic load Q_2

$$Q_2 = Q_1 (1 + 2)$$

This latter equation has been expanded by Eckenfelder and O'Connor (1961) and Eckenfelder (1963) to cover the effect of the decreasing removal of BOD per unit depth with increasing filter depth, i.e., residence time:

$$\frac{Le}{Lo} = e^{(-KD^{1-m})/(Q/A)^n}$$ (7)

where   Lo is the BOD of influent

D is the depth

1-m is the exponent of D, equal to 1.0 for a filter where biological slime is uniform with depth

If a waste is so complex that there is a decrease in the ratio of removal of BOD with time, the equation becomes

$$\frac{Le}{Lo} = \frac{1}{1 + \dfrac{CD^{(1-m)}}{(Q/A)^n}}$$ (8)

Certain values have been established on filter performance: C = 2.5, 1-m = 0.67, n = 0.5 for A in acres. D is the depth in feet and Q is the feed rate in 10^6 gal/day. Recirculation is treated strictly as a diluent of the influent feed Li and is expressed by

$$Lo = \frac{Li + RLe}{1 + R}$$

which was also described by Howland (1958). Galler and Gotaas (1964) used multiple regression analysis on data from existing plants and developed the expression

$$Le = \frac{0.464 \, Lo^{1.19} (1 + R)^{0.28} (Q/A)^{0.13}}{(1 + D)^{0.67} T^{0.15}}$$ (9)

The recirculation variable is included in this equation; however, Eq. (9) should be substituted for the applied BOD, i.e., Lo for filters using recirculation.

To obtain volumes by means of these three primary design approaches, the following formulas are used:

NRC: first stage

$$V = 0.0263 \ QLi \ \frac{(1 + 0.1R)^2}{1 + R} \ (\frac{E}{1 - E})^2 \tag{10}$$

second stage

$$V = 0.0263 \ \frac{QLi}{F} \ \left(\frac{E_2}{(1 - E_1)(1 - E_2)}\right)^2$$

Eckenfelder: first stage

$$V = 7.0 \ \frac{Q}{D^{0.33}} \ \left(\frac{\frac{E}{1 - E}}{1 + R}\right)^2 \tag{11}$$

second stage

$$V = 7.0 \ \frac{Q}{D^{0.33}} \ \left(\frac{E_2}{(1-E_1)(1-E_2)(1+R)}\right)^2$$

Galler-Gotaas: first stage

$$V = 0.1355D \left(\frac{Q^{0.13} \ Li^{0.19} \ \left[1 + R(1 - E)\right]^{1.19}}{T^{0.15} \ (1+D)^{0.67} \ (1-E)(1+R)^{0.78}} \Big/ (1-E_1)^{0.5}\right) \tag{12}$$

second stage

$$V = 0.1355D \left(\frac{Q^{0.13} Li^{0.19} \left[1 + R(1 - E) \right]^{1.19}}{T^{0.15}(1+D)^{0.67}(1-E)(1+R)^{0.78}(1-E_1)^{0.5}} \right)^8$$

A comparison of the variables involved in each of the formulas is best made in tabular form (Table 3). The trickling filter volume compared by the NRC and Eckenfelder formulas will vary directly with flow rate and with the efficiency functions $\dfrac{E}{(1-E)^2}$. Volume will decrease with increases in recirculation, reaching practical limiting recirculation ratios of 4-5. Volume varies with influent BOD in the NRC formula but is not a parameter in the Eckenfelder formula. Also in the Eckenfelder formula volume varies with the depth function $1/D^{0.33}$, which causes volume to decrease with depth. In the Galler-Gotaas approach, volume increases with flow rate, influent BOD, and required efficiency. It decreases with increases in temperature and depth. Recirculation and efficiency are independent parameters.

A volume decrease is shown only with higher efficiencies at practical limiting ratios of 4 or 5. Two-stage trickling filters can be designed by all three formulas. The optimal volume can be approximated by designing stages of equal volume with the NRC and Eckenfelder formulas, and using a ratio of 1:2 in the Galler-Gotaas formula. A two-stage unit requires less volume than a single stage unit.

In the future, basic models will apply a more mechanistic approach to trickling filter operation (Atkinson et al., 1963). Such models will probably treat the process with respect to fluid flow over a surface and develop an approach that permits examination of laminar or turbulent flow, steady state conditions, reaction rates, type of kinetics, solubility problems, interface studies, film thickness, and recirculation problems.

A. Design Cost

Optimization analysis for the design of trickling filters is presented in considerable detail by Galler and Gotaas (1966). A linear programming pro-

TABLE 3

COMPARISON OF DESIGN VARIABLE IN NRC, ECKENFELDER AND GALLER-GOTAAS FORMULAS

	Volume Function						Organic loading	Hydraulic loadings
	Q	BOD_i	D	R	T	Recycle ratio	lb/1000 ft^3	gal/day/ft^2
NRC	+	+	-	$\dfrac{(1+0.1R)^2}{1+R}$	-	4-8	9-196	34-734
Eck	+	-	$\dfrac{1}{D^{0.33}}$	$\dfrac{1}{(1+R)^2}$	+	4-5	9-196	34-734
G-G	+	+	+	+^{a}	+	4-5	30-110	230-690

[a] Recirculation and efficiency are interdependent.

cedure was developed for a trickling filter in which radius, depth, recirculation, feed rate, and BOD levels were considered. Primary and secondary settlers of rectangular design were included. Both design and cost calculations were presented. It was also found that depth and recirculation ratio were dependent on the BOD in the filter influent and the desired BOD reduction, and were independent of the plant influent volumetric rate. The hydraulic rate through a filter should be set at the maximum level permitted until the recirculation ratio is approximately 4 to 1. After this ratio is achieved, calculations show that the hydraulic rate should be decreased. Where pumping of influent is not a factor, trickling filters should be designed as deep as possible while maintaining adequate air circulation. Deep filters with forced air circulation are more economical than shallow filters in reducing BOD. However, they lose their advantage if pumping is required.

B. Recycling

Although recycling is commonly practiced in conventional systems, there is no single reason for justifying its use. Some of the factors involved are given here:

1. Valuable inorganic nutrients are brought to the site of maximum growth again.
2. It gives the recycle effluent a chance to pick up extracellular enzymes.
3. It dilutes the feed, lowering the level of any inhibiting compound present in the influent.
4. It often helps increase the hydraulic flow over the packing, thus ensuring adequate scouring.
5. It continually re-inoculates the system with a balanced set of microbial flora.

C. Oxygen Absorption in Trickling Filters

Oxygen absorbed by a liquid passing through a filter follows the relationship:

$$- dC/dD = K (C_s - C_e)$$

where $C_s - C_e$ is the oxygen deficit

D is the filter depth

K is the transfer rate coefficient

Integrating, we get

$$\frac{(C_s - C_e)_1}{(C_s - C_e)_2} = e^{K(D_2 - D_1)}$$

The transfer rate coefficient K is related to the hydraulic loading Q by the relationship:

$$K = C/Q^n$$

where   n depends on filter medium

C is the coefficient depending on filter medium

(See Table 2 for values of n and C.)

Oxygen transfer rates are expressed in lb $O_2/hr/ft^2/ft$ (mg $O_2/cm^2/m$) depth. Oxygen transferred to a typical polygrid filter is

$$N = 5.0 \times 10^{-4} K(C_s - C_e)/Q$$

where   N is lb $O_2/hr/ft^2/ft$

C_s is the oxygen sat., mg/liter

C_e is the DO concentration, mg/liter

Q is the hydraulic loading, gal/min/ft^2

Some typical oxygen utilization rates for filter slimes are presented in Table 4.

D. Forced Draft Ventilation

Forced draft ventilation of conventional trickling filters used in treating municipal sewage has little effect on effluent quality. It could have value if

TABLE 4

AVERAGE OXYGEN UTILIZATION RATE CHARACTERISTICS OF FILTER SLIMES[a]

Organic load (lb BOD/1000 ft³)	Hydraulic load[c] gal/min/ft³	Temp. (°C)	Slime[d] gal/ft³	O_2 uptake (mg O_2/hr/cm²)	Range of uptake rate
Spaced Polygrid					
534	3.0	30	113.4	0.0280	(0.0390 and 0.0165)
Polygrid					
184	0.75	24	54.1	0.0283	(0.0382 and 0.0235)
288	1.5	24	48.5	0.0186	(0.0235 and 0.0152)
452	3.0	30	220.0	0.0434	(0.0830 and 0.0190)

[a] After Eckenfelder and Barnhart, 1963.

[b] Multiply by 16.6 to obtain g/day/m³.

[c] Multiply by 40.7 to obtain liter/min/m².

[d] Divide by 0.028 to obtain gal/m³.

improved air distribution could be supplied. Its greatest application is where
the sewage temperature is high or in deep trickling filters. Under these
situations, the DO of the sewage is low initially and a good oxygen exchange is
required (Rhame et al., 1958). The rate of ventilation through the base of
a trickling filter is dependent largely on the temperature difference between
the filter and the atmosphere, and on the direction and the velocity of the wind.
Drafts due to air density are largely a function of temperature differences.
Humidity has very little affect on ventilation. The minimal ventilation rates
that have been tested are far greater than those theoretically required. Higher
drafts, but lower rates of air flow, generally occur in housed units, due to
the additional head losses imposed by the housing. Plastic media have been
used in bio-oxidation towers (Tow, 1960).

Actual draft through a filter has been measured by Johnson (1952), and
plots of bulk air flow against actual draft in inches give

$$h = 0.070V^{1.83}$$

$$h_m = 0.111 \times 10^{-3}V$$

$$V = 0.0519D$$

where h is the total draft or head (inches of water)

V is the bulk air flow ($ft^3/min/ft^2$ filter)

h_m is the draft or head loss through filter (per ft depth in inches
of water)

D is the sewage temp. minus outside air temp. in ^oF

E. Fluid Flow Through Trickling Filters

When packing media are small, the effect on residence time on the capillary water may be as large as 100%, especially when the rate of liquid application is low. On stone surfaces the flow is usually well distributed if the
surface feed is well dispersed. Consequently, channeling causes "n" (the
power to which flow is raised) to decrease less than two-thirds.

A comparison of the capillary effect of large and small stones shows that
high capillarity increases the values of "n" greater than two-thirds. Thus,
channeling and capillarity can compensate for each other and at standard
feed rates void space between stones will not be filled and "n" will be in a
region close to two-thirds. If flow becomes turbulent through the filter
bed rather than laminar, flow varies inversely with the one-third power of the
rate of flow.

A method of calculating capillary water is given by Howland et al. (1961-
63). In determining residence time, one must consider flow around spheres
(Sinkoff et al., 1959) plus the time required to pass through capillary water
bound between spheres, which is the capillary water divided by the rate of
flow.

$$\frac{tr}{H} = 0.0136 \left(\frac{S}{Q}\right)^{2/3} + \frac{ndy}{2}\left(\frac{S}{Q}\right)$$

or

$$\frac{tr}{H} = \frac{1.3(3\gamma)^{0.33}}{g^{0.33}}\left(\frac{S}{Q}\right)^{0.67} + \frac{ndy}{2}\left(\frac{S}{Q}\right)$$

                    flow around sphere   flow through capillary

Here, tr is the residence time in sec, H is the filter depth in ft, S is π/d the
specific surface, d is the diameter of the sphere in feet, Q is the rate of flow
in $ft^3/sec/ft^2$ of filter, y is the meniscus volume function (vol. of meniscus +
πd^3), n is the number of meniscus entached per sphere, γ is the viscosity
$(1.23 \times 10^{-5} ft^2/sec)$, and g is $32.2 ft/sec^2$.

The capillary contacts between stones behave like channels when liquid
flows through them, thus making them act like partially filled cylindrical
tubes. Howland et al. (1961-1963) have given good kinetic treatment of this
consideration.

Residence time characteristics for several types of packing media are
presented in Table 5. Residence time is a function of hydraulic loading,
particle size, and depth.

Hydraulic loadings of 203×10^6 gal/acre/day with BOD loading varying

TABLE 5

RESIDENCE CHARACTERISTICS IN THE PRESENCE OF FILTER SLIME[a]

Media	Hydraulic Loading[b] (gal/min/ft^2)	Residence time (min)	Depth[c] (ft)
Rock--3/4 to 1 1/2 in. (1.9 to 3.8 cm)	0.093	35	6.0
Rock--1 1/2 to 3 in. (3.8 to 7.6 cm)	0.093	26	6.0
Polygrid	1.0	6	18.0
Asbestos	1.0	12.2	18.0
Asbestos	5.0	3.3	18.0
Polygrid[d]	2.0	3.8	18.0

[a] After Eckenfelder and Barnhard, 1963.

[b] Multiply by 40.7 to obtain liter/sec/m^2.

[c] Multiply by 0.3 to obtain m.

[d] Thin slime cover

varying from 200-750 ppm have been tested on polygrid filters (Fig. 4). The
present reduction of BOD is independent of the level of BOD in the feed (Egan
and Sandlin, 1960). Although the efficiency of BOD removal is not great, it
is about 42% over a loading range of 200-750 lb of BOD/1000 ft^3 of filter per
day. Removal of up to 370 lb of BOD/1000 ft^3/day may be obtained at hydraulic
loads exceeding 175×10^6 gal/acre/day.

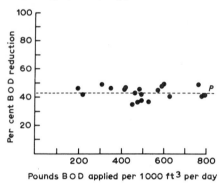

FIG. 4. The relationship between percent reduction and BOD applied.

Even with all the current information available, there are still large
gaps in the data on the true effects of temperature, film thickness, sloughing,
hydraulic loadings, organic loadings, mass and nature of the biological film
transfer, absorption of oxygen contribution from extra-cellular enzymes, and
the required contact time between organic matter and biological film. Considerably more effort must be made into the investigation of forced aeration,
particularly pertaining to high BOD loadings and deep filters. It is also
surprising that little work has been done on recycling slimes from the secondary clarifier to the primary clarifier to aid in flocculation and sedimentation.
A partial activated sludge system coupled with a trickling filter may have
definite areas of application.

REFERENCES

Atkinson, A., A. W. Busch, and G. S. Dawkins, 1963. "Recirculation,
reaction kinetics, and effluent quality in a trickling filter flow model." J.
Water Pollution Control Fed. 35, 1307-1317.

Baker, J. M. and Q. B. Graves. 1968. "Recent approaches for trickling
filter design." Proc. Am. Soc. Civ. Eng. (Sanit. Div.) 94, 65-85.

Eckenfelder, Jr., W. W. 1963. "Trickling filtration design and performance." Trans. A.S.C.E. 128, 371-398.

Eckenfelder, Jr., W. W. and E. L. Barnhart. 1963. "Performance of a high rate trickling filter using selected media." J. Water. Pollution Control Fed. 35, 1535-1551.

Eckenfelder, Jr., W. W. and D. J. O'Connor. 1961. Biological Waste Treatment. Pergamon, New York.

Egan, J. T. and M. Sandlin. 1960. "Evaluation of plastic trickling filter media." Ind. Wastes. 5, 70-77.

Galler, W. S. and H. B. Gotaas. 1964. "Analysis of biological filter variables." J. Sanit. Eng. Div., Proc. A.S.C.E. Paper No. 4174, 90, 59-79.

Galler, W. S. and H. B. Gotaas. 1966. "Optimization analysis for biological filter design." J. Sanit. Eng. Div. 92, 163-182.

Howland, W. E. 1953. "Effect of temperature on sewage treatment processes." Sewage Ind. Waste 25, 161.

Howland, W. E. 1958. "Flow over porous media as in a trickling filter." Proc. 12th Ind. Waste Conf., Purdue Univ. Ext. Ser. 94, 435-465.

Howland, W. E., F. G. Pohland, and D. E. Bloodgood. 1961-63. "Kinetic in trickling filters." Air and Water Pollution, 5, 233-248.

Johnson, W. K. 1952. "Ventilation of trickling filters." Sewage Ind. Waste 24, 135-148.

Minch, V. A., J. T. Egan, and M. Sandlin. 1962. "Design and operation of plastic filter media." J. Water Pollution Contr. Fed. 34, 459-469.

Munteanu, A., E. Cute, S. Godeanu, A. Eminovice and C. Murgoci. 1967. "Treatment of waste waters from synthetic rubber industry." Studii de. Protectia si Epurarea Apelor. 9, 123-163.

National Research Council, 1946. "Sewage treatment at military operations." Rept. Comm. Sewage Treatment on Sanitary Eng. NRC, Div. Med. Sci., Wash. D. C.

Rhame, G. A., C. R. Beoll, C. Doyle, and E. Williams. 1958.
"Effects of forced draft ventilation at a municipal sewage treatment plant."
Sewage Ind. Waste. 30, 1308-1311

Schulze, K. L. 1960a. "Elements of a trickling filter theory."
Proc. 3rd Biol. Waste Treatment Conf. Manhattan College.

Schulze, K. L. 1960b. "Load and efficiency of trickling filters." J.
Water Pollution Contr. Fed. 32, 245-261.

Sinkoff, M. D., R. Porges and J. H. McDermott. 1959. "Mean
residence time of a liquid in a trickling filter." J. Sanit. Eng. Div. A.S.C.E.
85 (No SAb), p. 51.

Tow, D. J. 1960. "Application of the plastics polygrid packing to
bio-oxidation towers." Ind. Chem. 60, 265-270.

Velz, C. J. 1948. "A basic law for the performance of biological filters."
Sewage Works J. 20, 607-617.

Chapter 9

ANAEROBIC DIGESTION

Anaerobic digestion of waste is not a simple process. It is more complex than most scientific articles on the subject indicate. As with many bio-oriented processes the art has been practiced for centuries but the science is still forthcoming.

In operation, either a low or high solids waste is pumped into a reservoir, where anaerobic conditions prevail, and the anaerobic microbes are permitted to reduce the level of the waste. The primary end products if properly operated are methane, hydrogen, and carbon dioxide, with hydrogen sulfide being a primary product when operated improperly. Environmental conditions influence operations greatly and there are at least two ecological groups of microbes that participate in the process: (a) organic acid (volatile acids) and (b) methane-producing bacteria (Fig. 1).

Historically, combustible gases produced under natural conditions were reported by Volta in 1776. This process accounts for the vast quantities of natural gas that have been mined and that exist in nature.

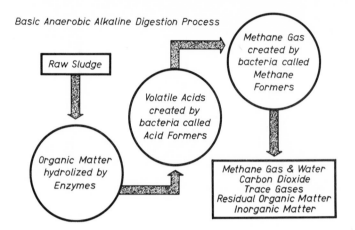

FIG. 1. Process by which anaerobic microbes remove organic material from waste.

I. MICROBIOLOGY AND BIOCHEMISTRY

To understand anaerobic digestion it is necessary to become acquainted with the microbes and the biochemical pathways involved. Organic wastes are usually high in polysaccharide carbohydrates, fats (or oils), and proteins. These high molecular weight compounds are degraded catalytically to smaller molecules (Fig. 2) by enzynes from a diverse and complex microbial population. Proteins are biodegraded to amino acids, imino acids, and ketoacids, and polysaccharides to glucose and eventually pyruvic acid and other organic acids. Pyruvic acid is the cynosure in this metabolic maze, and from it three carbon keto-carboxylic acid microbes synthesize a diverse number of intermediates.

Of these, the volatile organic acids and alcohols are of prime interest. (Fig. 3 a, b). Some of the alcohols formed are ethanol, acetymethyl carbanol, 2,3-butanediol, butanol, and isopropanol. Organic acids synthesized are lactate, oxalacetate, succinate, propionate, acetolactate, butyrate, acetate, and formate. Most of these compounds are synthesized by members of the clostridia and coli-aerogenes groups of bacteria. This list should also include all the di- and tricarboxylic acids (TCA) of the Krebs TCA-cycle. Volatile acids are also synthesized from amino acids without being converted to pyruvic acid. Leucine and valine are oxidized via the Strickland reaction to isovaleric and isobutyric acids. Glycine and proline and other amino acids

DIGESTION anaerobic —— *optimum operating state*

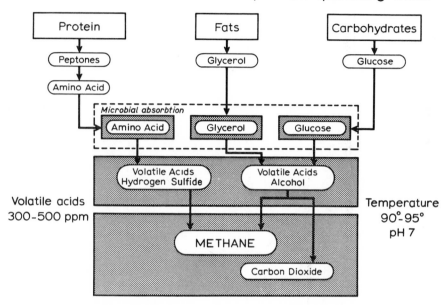

FIG. 2. Anaerobic bioconversion of protein, fats, and carbohydrates to methane and carbon dioxide.

act as acceptors for the hydrogen produced. Isovaleric acid is found in digesting sewage sludge (Liubimov and Kagan, 1958).

$$CH_3-CH-CH_2-\overset{\overset{\displaystyle NH_2}{|}}{\underset{\underset{\displaystyle H}{|}}{C}}-COOH +2H_2O \rightarrow CH_3-\underset{\underset{\displaystyle CH_3}{|}}{CH}-CH_2-COOH +NH_3$$
$$+4H +CO_2$$

$$\text{Leucine} \qquad\qquad\qquad \text{Isovaleric Acid}$$

$$CH_3-\underset{\underset{\displaystyle CH_3}{|}}{CH}-\underset{\underset{\displaystyle H}{|}}{\overset{\overset{\displaystyle NH_2}{\diagdown}}{C}}-COOH +2H_2O \rightarrow CH_3-\underset{\underset{\displaystyle CH_3}{|}}{CH}-COOH +NH_3 +4H +CO_2$$

$$\text{Valine} \qquad\qquad\qquad \text{Isobutyric Acid}$$

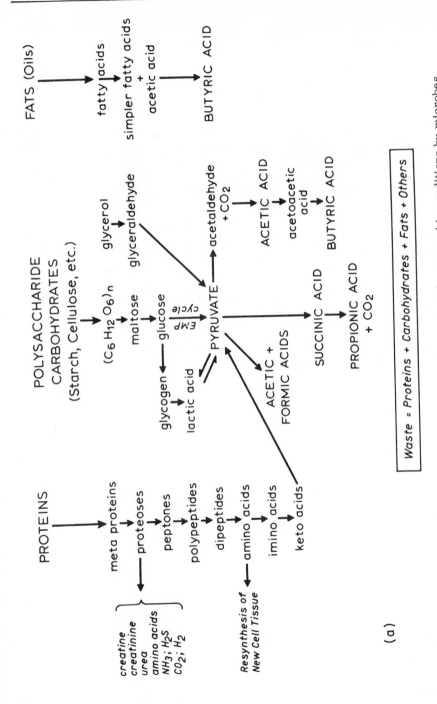

Fig. 3(a). Biosynthesis of pyruvate and other intermediates under anaerobic conditions by microbes.

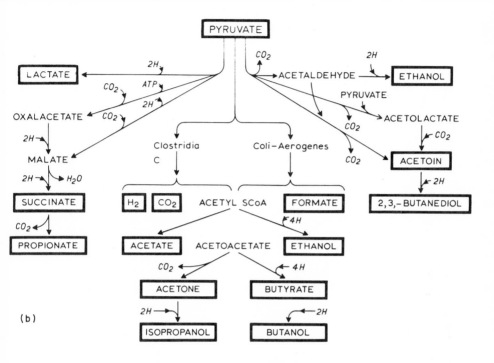

FIG. 3(b). Bioconversion of pyruvate to other organic intermediates.

Some of the more common bacteria synthesizing organic acids are summarized in Table 1.

Raw sewage contains about 45 mg/liter of fat, which is always high in palmitic and oleic acids. Fats are hydrolyzed to glycerol and long chain fatty acids such as stearic, oleic, palmitic, etc. Palmitic and octanoic acids are degraded by β-oxidation to acetic acid, CO_2, and methane (Jeris and McCarty, 1962). Long chain paraffinic hydrocarbons are oxidized by monoterminal or diterminal oxidation to a long chain fatty acid. (Kester and Foster, 1963) and oxidized further by β-oxidation.

The type of microbial population varies as the waste substrate changes. In the presence of cellulosic waste, Bacillus cereus, B. megaterium, Alcaligenes faecalis, Proteus vulgaris, Pseudomonas aeruginosa, Ps riboflavina, Ps reptilovora, and Leptospira biflexa have been isolated from anaerobic

TABLE 1

ORGANIC ACID PRODUCING MICROBES

Microbe	pH	Temp. (°C)	Products		References
Bacillus cereus	5.2	25-35	acetic	lactic	Toecrien, 1967
Bacillus knelfelkampi	5.2-8.0	25-35	acetic	lactic	Burbank et al., 1966
Bacillus megaterium	5.2-7.5	28-35	acetic	lactic	Toecrien, 1967
Bacteriodes succinogenes	5.2-7.5	25-35	acetic	succinic	---
Clostridium carnofoetidum	5-8.5	25-37	formic, acetic,		Burbank et al., 1956
Clostridium cellobioparus	5-8.5	36-38	lactic, ethanol, CO_2		---
Clostridium dissolvens	5-8.5	35-51	formic, acetic, lactic		---
Clostridium thermocellulaseum	5-8.5	55-65	formic, acetic lactic, succinic		---
Pseudomonas formicans	---	33-42	formic, acetic, lactic, succinic, ethanol		Wood, 1961
Ruminocossus flavefaciens	---	33-38	formic, acetic, succinic		---

digesters (Toerien, 1967). Starch wastes are known to support growth of
Micrococcus candidus, M. varians, M. ureae, Bacillus cereus, B. megaterium, and many species of Pseudomonas. Wastes high in protein support
populations of Bacillus cereus, B. circulans, B. pumilis, B. sphaerius, B.
subtilis, Micrococcus varians, Escherichia coli, Paracolobactrum inter-
medium, P. coliforme, and species of Pseudomonas. Vegetable oils such
as sunflower oil encourage the growth of Micrococcus, Bacillus, Streptomy-
ces, Alcaligenes, and Pseudomonas (Torien, 1967) in anaerobic digesters.

Filamentous fungi appear to have very little function in anaerobic digestion although they may participate in organic acid and alcohol production.

Methane production in small amounts by anaerobic microorganisms may
be far more common than current literature would indicate. However, certain bacteria possess unique abilities to produce high levels of methane under
specific nutritive and ecological conditions. The bacteria responsible for
methane production belong to the following genera: Methanobactercum,
Methanobacillus, Methanosarcina, and Methanococcus (Table 2). The chemical reactions proposed for some of the substrates and microbes are sum-
maries in Table 3.

Biomethane is produced in the following anaerobic environments:

1. Carbon dioxide and hydrogen

$$CO_2 + 4H_2 \longrightarrow CH_4 + 2H_2O$$

2. Formic acid

$$4HCOOH \longrightarrow CH_4 + 3CO_2 + 2H_2O$$

3. Acetic Acid

$$CH_3COOH \longrightarrow CH_4 + CO_2$$

4. Ethanol

$$2CH_3CH_2OH \longrightarrow 3CH_4 + CO_2$$

The general reaction is regarded as proceeding in two steps, i.e. cleav-
age of H_2O and reduction of CO_2:

$$CH_3COOH + H_2O \longrightarrow 2CO_2 + 8(H)$$

$$8(H) + CO_2 \longrightarrow CH_4 + 2H_2O$$

TABLE 2

THE METHANE BACTERIA [a,b]

ORGANISMS	pH	Temp.	ACIDS METABOLIZED
Methanobacterium omelianskii	6.5–8	37–40°C	CO_2, H_2, ethanol, primary and secondary alcohol
Methanobacterium propionicum			Propionate
Methanobacterium formicum			H_2, CO_2, formate
Methanobacterium suboxydans			Butyrate, valerate, caproate
Methanobacterium sohngenii			Acetate, butyrate
Methanobacterium ruminantium			
Methanobacterium soehngenii			Acetate, formate
Methanococcus vannielli	7.4–9		Formate
Methanococcus mazei		30–37°C	Acetate, butyrate
Methanosarcina methanica		35–37°C	Acetate, butyrate
Methanosarcina barkerii	7.0	30°C	CO_2, H_2, acetate, methanol

[a] In general, the methane bacteria will tolerate a pH range of 6.5–7.4.

[b] After Barker and Burbank (1966).

TABLE 3

METHANE FERMENTATION REACTIONS

ORGANISMS	REACTIONS
Methanobacterium soehngenii Methanococcus mazei Methanosarcina methanica Methanosarcina barkeri	$CH_3COOH \longrightarrow CH_4 + CO_2$
Methanobacterium propionicum Methanococcus mazei Methanosarcina methanica Methanobacterium suboxydans	$4CH_3CH_2COOH + 2H_2O \rightarrow 7CH_4 + 5CO_2$ $2CH_3(CH_2)_2COOH + 2H_2O \longrightarrow 5CH_4 +$ $3CO_2$ $^a 2CH_3(CH_2)_2COOH + 2H_2O + CO_2 \longrightarrow$ $CH_4 + 4CH_3COOH$
Methanobacterium omelianskii	$2CH_3CH_2OH \longrightarrow 3CH_4 + CO_2$ $2CH_3CH_2OH + CO_2 \rightarrow CH_4 +$ $2CH_3COOH$
Methanobacterium suboxydans	$CH_3COCH_3 + H_2O \qquad 2CH_4 + CO_2$
	Organic acid reaction (β-oxidation) $CH_3(CH_2)_{16}COOH + 16H_2O \longrightarrow$ $9CH_3COOH + 3 \cdot 16H_2$ Methane synthesis $4CO_2 + 16H_2 \longrightarrow 4CH_4 + 8H_2O$ $9CH_3COOH \longrightarrow 9CH_4 + 9CO_2$
	$CH_3(CH_2)_{16}COOH + 8H_2O \longrightarrow 13CH_4 +$ $5CO_2$

^{a}Organism not identified.

This two step process is almost identical to certain autotropic Carboxy-domonas species, which oxidize carbon monoxide to carbon dioxide under anaerobic conditions:

$$CO + H_2O \longrightarrow CO_2 + H_2$$

The CO_2 and H_2 are then converted to methane:

$$2CO_2 + 2H_2O \longrightarrow CH_4 + CO_2$$

It appears that high levels of volatile organic acids (2000 mg/liter) as well as high levels of sodium ion decrease the activity of methane bacteria (McCarty and McKinney, 1961). This is particularly true if the salt is added on a slug basis. The gas composition shows some variation, but the levels of methane should always exceed 60% (Table 4). Other major gases present are H_2, CO_2, and N_2. Growth yields of the methane bacteria on methanol, formic acid, acetic acid, and propionic acids (Table 5) have been reported by McCarty (1966). The highest rate of substrate removal is for formic acid (22 g/day/g bacteria); however, the highest growth yield is observed with methanol (0.16 g bacteria/gram substrate).

There appears to be a close relation between the biochemical function of corrinoids and methane synthesis. Barker (1967) extracted abundant corri-noids from extracts of Methanobacillus omeliansky and Methanosarchina bar-keri. Both folate and B_{12} derivatives were shown schematically to be involved in methane biosynthesis from CO_2, formate, serine, methanol, acetate, acetone-butanol (Pantskhava and Bykhovskii, 1965).

TABLE 4

GAS COMPOSITION FROM ANAEROBIC DIGESTER[a]

Gas composition	%
CO_2	25 – 35
H_2	1.0 – 5
CH_4	50 – 68
N_2	2.0 – 7
O_2	0.0 – 0.1

[a]After Langford, 1957

TABLE 5

METHANE FERMENTATION OF DIFFERENT SUBSTRATES[a]

Substrate	Max. rate of substrate use (g/day/g bacteria)	Growth yield (g/bacteria/g substrate)
Methanol	3.1	0.16
Formic acid	22.0	0.022
Acetic acid	2.1	0.073
Propionic acid	3.7	0.045

[a]After McCarty (1966).

II. KINETICS

Anaerobic digesters contain a heterogenous mixture of solids, liquids, and gases. The solids contain inert inorganics as well as biodegradable organics and a complex microbial population. The microbes utilize the solids (dissolved and particulate) to generate energy and synthesize cell material. The trophic level (BOD) is reduced and the end products of CO_2 and H_2O increase. The manner in which the BOD is decreased is accounted for by two theories: the surface reaction theory and the enzyme catalyzed theory.

A. Surface Reaction Theory

This surface reaction theory is based on the hypothesis of Langmuir, which assumes (a) the layer of substrate molecules absorbed by the active microbial mass is only one molecule thick and (b) at the microbe surface there is a constant condensing and evaporation of substrate molecules.

Let θ be the fraction of microbe surface covered by substrate at any instant of time and

s be substrate concentration at any time.

Then r_c (rate of condensation of substrate molecule)

$$= k_1 (1 - \theta) s$$

where k is a constant. Similarly

where r_e is the rate of evaporation of substrate from the microbe surface and k_2 is a proportionality constant

$$r_e = k_2 \theta$$

Assuming an equilibrium state where the rates of condensation and evaporation are equal, we obtain

$$r_c = r_e$$

$$k_1 (1-\theta) s = k_2 \theta$$

$$\theta = \frac{s}{k_2/k_1 + s}$$

Setting $y = k_2/k_1$, then

$$\theta = \frac{s}{y+s} \tag{1}$$

Considering the utilization of the substrate molecules condensed on the microbe's surface, and the surface as the active region, then the rate of substrate utilization per unit weight of active solid is expressed by

$$M = K \tag{2}$$

where K = a constant accounting for active surface area per unit weight of microbes. The substrate concentration per unit area of active surface, and the removal of substrate from the active surface for the microbial system, are used to express the substrate utilization rate as a function of substrate concentration [substitute Eq. (1) in Eq. (2)] :

$$M = K \left(\frac{s}{y+s} \right)$$

B. Enzyme Catalyzed Reaction Theory

This theory is based upon microbial growth and the production of intra- and extracellular enzymes. Enzymes are known to be highly specific for specified substrates. They react to form an enzyme-substrate complex (Michaelis-Menten), which dissociates to give a product and a free enzyme.

$$\text{Enzyme (E) + Substrate (S)} \left(\frac{k_2}{k_1} \right) \text{(ES)} \xrightarrow{k_3} \text{E + Product} \tag{4}$$

The rate of change of enzyme-substrate complex may be expressed in differential form.

$$\frac{d\,(ES)}{dt} = k_1\,(E - ES)S - k_2\,(ES) - k_3\,(ES) \tag{5}$$

where   E is the enzyme

S is the substrate

ES is the enzyme-substrate complex

$k_1, k_2, k_3,$ are the reaction rate constants

At steady state there will be no net change in the ES complex; thus

$$\frac{d\,(ES)}{dt} = 0$$

The above formula can then be solved for ES, which yields

$$ES = \frac{(S)k_1\,(E)}{Sk_1 + k_2 + k_3} = \frac{(S)\,(E)}{\left[\dfrac{(k_2 + k_3)}{k_1}\right] + S} \tag{6}$$

This is the common form of the Michaelis-Menten equation. The rate of product formation is highly dependent upon the dissociation of (ES) and k_3:

$$(ES) \xrightarrow{k_3} E + P$$

The rate of product formation here is

$$\nu = k_3\,(ES) \tag{7}$$

or

$$\nu = \frac{k_3\,(E)\,(S)}{\left(\dfrac{k_2 k_3}{k_1}\right) + S} \tag{8}$$

Many times the substrate far exceeds the concentration of enzymes, which results in most enzymes at any particular time being complexed. The maximum production rate is

$$\nu\,max = k_3\,E$$

The rate of product formation becomes

$$\nu = \frac{V_{max}\,S}{\left[\dfrac{(k_2 + k_3)}{k_1}\right] + S} \tag{9}$$

where, by definition

$$K_m = \frac{(k_2 + k_3)}{k_1}$$

giving the standard form of the Michaelis-Menten equation

$$\nu = \frac{V_{max}\,S}{Km + S} \tag{10}$$

One chould note the similarity in the equations for the surface reaction theory and the enzyme catalyzed models.

C. Continuous Flow Model

Enzyme formulation is dependent on microbial growth, and the operation of a single stage anaerobic digester is usually fed intermittently or continuously. This requires analysis of the system in terms of continuous flow fermentation. The most simple to analyze is a completely mixed single stage reactor; however, mathematical treatment of pure complex systems are available (Herbert, 1961). The theory of continous flow models for growth of bacteria is described by Novick and Szilard (1950) and Monod (1950). This development has permitted application to bacterial mutation (Moser, 1958), rumen metabolism (Quinn, 1962), lactic acid (Leudeking and Piret, 1959), aerobic stabilization of wastes (Gram, 1956), anaerobic stabilization of wastes (Andrews and Pearson, 1965), and algal production by the Carnegie Institute of Washington (1953). Malek (1961) and Malek and Beran (1962) have given quite detailed mathematical treatment of microbial growth.

The kinetic equations for continuous flow systems are developed by applying material balances to three fundamental relationships: (a) growth rate, (b) essential nutrient (or waste substrate) and growth rate, and (c) growth yield. The logarithmic growth rate of bacteria is expressed in the form of a common autocatalytic equation:

$$\frac{dx^\circ}{d\theta} = kX^\circ \tag{11}$$

Where x° is the microbe concentration (mass/vol)

θ is the time

k   is the specific growth rate (time $^{-1}$) and is always determined in excess substrate

Bacterial concentrations are expressed in terms of cells/ml, but it is often more convenient to express in mass/unit volume. Although the two expressions are not equivalent under all conditions with complex microbial populations, the latter is required. In anaerobic digestion, decay of cells by autolysis and lysis is as important as growth. This must be taken into account by utilizing a specific decay rate factor:

$$\frac{dX^\circ}{d\theta} = (k - k^d)\ X^\circ \tag{12}$$

where K^d = specific microbial decay rate (time $^{-1}$).

The importance of k_d is really not known; however, since residence times in digesters are usually greater than 12 days, the term may have greater significance as more information is developed. Growth rate constants are always determined in excess substrate. Limiting essential nutrients limits microbial growth.

The relationship between substrate and growth rate follows a hyperbolic function, which is quite similar to the Michaelis-Menten equation describing enzyme-substrate interaction. This correlation is not strictly fortuitous since substrate adsorption, transport, and the enzymatic conversion of essential nutrients all fit into the same general category of ractions. This is described in the Monod equation by

$$k = \frac{k^m X^s}{K^s + X^s} \tag{13}$$

Where k^m is the maximum growth rate, time^{-1}

X^s is the conc. essential nutrient, mass/vol.

K^s is the substrate conc. at 1/2 maximum growth rate (mass/vol)
       --a constant

The growth rate of an individual microbe or a complex microbial population can be controlled by a wide variety of inorganic or organic compounds essential for life processes (James, 1961). These may be (a) essential amino acids; (b) energy sources such as carbohydrate, fats, oils; (c) nitrogen sources such as ammonia, nitrate, and urea; and (d) mineral elements such as phosphate, sulfate, iron, or trace minerals.

Of course, many other factors limit growth and have been used in growth studies, e.g. population density per se, pH, toxic products, inhibitors, environmental factors, etc. The Monod expression of growth not only is used when a single compound controls the growth rate, but has found application involving complex media and in mixed cultures where the nature of the factors limiting growth are not known. Growth studies show that for a given microbe and a given set of nutritive conditions, the weight of nutrient utilized is a remarkable constant value. This is expressed as $Y°$, which is a dimensionless number:

$$Y° = \frac{\text{mass of microbes}}{\text{mass of essential substrate utilized}} \tag{14}$$

The rate of change of cell mass is expressed in differential form by

$$\frac{dX^\circ}{d\theta} = Y^\circ \frac{dX^s}{d\theta}, \quad \text{or} \quad \frac{dX^\circ}{dXs} = -Y^\circ \tag{15}$$

This shows the relationship of the growth yield constant to growth rate. Notably Y° does change if the population produces large amounts of cell storage products, polysaccharide, poly-β-hydroxybutyric acid, etc.

D. Single-Stage Digesters

Since most anaerobic digesters can be treated as completely mixed single-stage reactors, a mathematical model is developed for such a unit. A number of basic assumptions are made (Herbert et al., 1956):

1. A constant proportion of the organisms are viable.
2. The primary substrate concentration serves as the essential or growth limiting substrate.
3. The Monod equation is applicable.
4. The mass of organisms produced per mass of substrate utilized is constant.

A material balance to handle organisms, substrate, products, and other components of the system is needed   (Fig. 4).

$$\begin{bmatrix} \text{Change in} \\ \text{reactor} \end{bmatrix} = \begin{bmatrix} \text{Organisms} \\ \text{in influent} \end{bmatrix} + \begin{bmatrix} \text{Growth} \end{bmatrix} - \begin{bmatrix} \text{Lysis} \\ \text{of} \\ \text{organisms} \end{bmatrix} - \begin{bmatrix} \text{Organisms} \\ \text{in} \\ \text{effluent} \end{bmatrix}$$

$$\frac{VdX^\circ_1}{d\theta} = X^\circ_o \, F + k\,X^\circ_1 V - k^d X_1^{\,\circ} V - X_1^{\,\circ} F \tag{16}$$

Where   V is the reactor volume

X°_o is the organism concn. in influent, mass/vol.

X°_1 is the organism concn. in reactor, mass/vol.

F is the flow rate, vol/time

Since the contribution of the organisms in the influent is negligible compared with those in the reactor, one can treat the influent as if no organisms were present. This might not be true for some types of wastes, but it is an assumption that applies under most conditions. The residence time for the

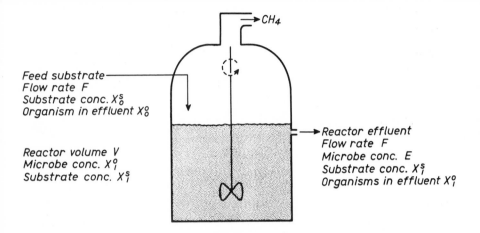

FIG. 4. Single-stage continuous flow reactor.

reactor is a function of reactor size and flow rate, i.e., $\theta^r = V/F$. Thus our ideal equation expressing changes in the material balance may be simplified and arranged as follows:

$$\frac{dX^o}{d\theta} = \left[k - k^d - \frac{1}{V/F} \right] X_1^o$$

or

$$\frac{dX^o}{d\theta} = \left[k - k^d - \frac{1}{\theta^r} \right] X_1^o \tag{17}$$

At steady state, i.e. when $\dfrac{dX^o}{d\theta} = 0$, the net specific growth rate of $(k - k^d)$ is equal to the reciprocal of the residence time:

$$k - k^d = \frac{1}{\theta^r} \tag{18}$$

This relationship is vital to understanding reactor kinetics because it shows how growth rate can be controlled by varying the flow rate in a constant volume reactor. The flow rate can be increased to the point in which dilution exceeds the growth of the microbial population. The critical dilution where this occurs is termed washout. Just before this is reached, a flow rate in which the output of microbes is maximum is usually observed. (Cantois, 1959; Schroepfer and Ziemke, 1953; Pfeffer et al., 1967).

A material balance must also be made to handle the concept of substrate or waste:

$$
\begin{bmatrix} \text{Substrate} \\ \text{change in} \\ \text{reactor} \end{bmatrix} = \begin{bmatrix} \text{Substrate} \\ \text{in} \\ \text{influent} \end{bmatrix} - \begin{bmatrix} \text{Substrate} \\ \text{consumed} \end{bmatrix} - \begin{bmatrix} \text{Substrate} \\ \text{in} \\ \text{effluent} \end{bmatrix}
$$

$$
\frac{V dX_1^S}{d\theta} = X_o^S F - \frac{k X_1^o V}{Y^o} - X_1^S F \qquad (19)
$$

Here, X_o^S = substrate concentration in the influent in mass/vol, X_1^S = substrate concentration in the effluent in mass/vol, and $k X_1^o V / Y^o$ is the substrate consumed by the organisms. This is obtained from the relationship

$$
\frac{d X^o}{d X^S} = -Y^o .
$$

Again taking the dilution rate $F/V = 1/\theta^r$ and the net change in specific growth rates, $k - k^d = 1/\theta^r$, these can be substituted into the generalized material balance equation to obtain

$$
\frac{d X_1^S}{d\theta} = \frac{1}{F/V} \left[X_o^S - X_1^S - \frac{X_1^o}{Y^o} \right] - \frac{X_1^o k^d}{Y^o}
$$

$$
\frac{d X_1^S}{d\theta} = \frac{1}{\theta^r} \left[X_o^S - X_1^S - \frac{X_1^o}{Y^o} \right] - \frac{X_1^o k^d}{Y^o} \qquad (20)
$$

As noted in the Monod equation Eq. (13), the specific growth rate k is a function of X_1^S. Thus k in eq. (17) is replaced by

$$
\frac{d X_1^o}{d\theta} = X_1^o \left[\frac{k^m X_1^S}{K^S + X_1^S} - kd \frac{1}{\theta^r} \right] \qquad (21)
$$

Under steady state operation, $\dfrac{d X_1^S}{d\theta}$ and $\dfrac{d Xi}{d\theta}$ of Eqs. (20) and (21), respectively, become zero. These equations can then be solved for

X_1^S and X_1^o, the substrate and cell mass concentrations under steady-state operation:

$$X_1^S = \frac{K^S(1 + k^d\theta^r)}{\theta^r k^m - (1 + k^d\theta^r)} \qquad (22)$$

$$X_1^o = Y^o\frac{(X_o^S - X_1^S)}{1 + k^d\theta^r} \qquad (23)$$

Before Eqs. (22) and (23) can be used, values for the constants k^m, k^d, K^S, and Y^o must be determined for the organism(s), substrate concentration in the reactor and effluent for any value of residence time θ^r or concentration of influent substrate X_o^S. These values may be obtained quite easily from experimental data by employing Lineweaver-Burk plots, i.e., the reciprocal of velocity is plotted against the reciprocal of substrate. This is often referred to as a double reciprocal plot. The yield constant Y^o and decay rate k^d are determined from a plot of the double reciprocal form of eq. (23):

$$\frac{X_o^S - X_1^S}{X_1^o} = \frac{k^d}{Y^o}\,\theta^r + 1/Y^o$$

whereas k^S and k^m are obtained by using the double reciprocal form of Eq. (22).

$$\frac{\theta^r}{1 + k^d\theta^r} = \frac{K^S}{km}\,X_1^S + 1/k^m$$

With all the assumptions made, this mathematic model works extremely well. It does not consider many factors such as mutations, changes in viability, gross changes in the organism population, and effect of toxic products, but in spite of deficiencies its validity is established.

Andrews and Pearson (1965) report a maximum methane-COD production rate of 0.81 g/g volatile solids (Table 6). With long residence times, where methane production predominates, a value for Y^o was 0.14 mg VSS/mg COD and k^d was 0.22 day^{-1}. At lower residence times, Y^o increases to 0.54 mg VSS/mg COD and k^d increases 0.87 day^{-1}. This emphasizes the need to include k^d in kinetic studies.

III. HISTORICAL BACKGROUND

Anaerobic digestion, a naturally occuring process that probably accounts for the millions of cubic feet of natural gas mined each year, was first recognized by a Frenchman, Louis Mouras, in 1960. Mouras observed the liquifaction of excrementitious matter in a closed vault with a water seal. In 1895,

TABLE 6

CONSTANTS FOR KINETIC MODELS

Constants	OA[a,b]	Methane Producing[b]	Packing House Waste[c]	Synthetic Milk Waste[d]
k (day^{-1})	1.33	< 1.33	0.24	0.13
k (day^{-1})	0.87	0.02	0.17	0.06
K^S (mg/liter COD)	43.7	40.0	5.5	22.3
$Y°$ (mg VSS/ mg COD)	0.54	0.14	0.76	0.38
Temp ($°C$)	38	38	36	20-25

[a] Organic acid producing.

[b] Andrews and Pearson, 1965.

[c] Schroepfer and Ziemke, 1959.

[d] Gates et al., 1967.

Donald Cameron designed a septic tank for the City of Exeter and it was reported that 80% reduction in solids was achieved; in addition, a combustible gas in sufficient quantities was produced and used for lighting the vicinity near the plant. In 1904, the Travis hydraulic tank, a dual purpose sedimentation-digestion tank, was installed. In 1907, Dr. Imhoff developed and designed the Imhoff tank, another dual purpose system. By the mid 1920's a move was made toward the separate sludge digestion tank. This was furthered by the invention of the Down's floating cover for tanks, which made it possible to fill and draw liquor from these digestion tanks without the hazard of explosions resulting from air mixing with the methane gas. With the advent of the separate tank system came the conventional process as practiced today. With the increasing treatment demands and technological development after 1950, the high rate unit and the anaerobic contact unit were developed.

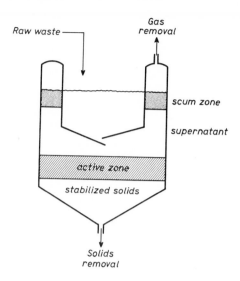

FIG. 5. Imhoff tank. Dual purpose sedimentation and anaerobic digestion system.

A. Imhoff Tank (Fig. 5)

This was designed for a dual purpose. It served as a settling tank for solids removal to an anaerobic chamber in which it was anaerobically decomposed. Since the anaerobic chamber was not mixed, there was stratification that is characteristic of the conventional process.

B. Conventional Process

In this process freshly introduced solids will shortly rise to the scum layer. As decomposition proceeds in this layer there is a constant shedding of the partially digested solids that fall to the bottom of the digester, forming a layer of digested sludge (Fig. 6). The scum layer constitutes a zone of primary decomposition where the majority of the gases are released. The sludge zone constitutes a secondary zone of decomposition, and the supernatant, rich in dissolved and suspended matter, can be transferred to other units for further treatment.

These units may operate well under conditions of low organic loadings and long detention time, 50 to 60 days, or more. Once the organic loadings

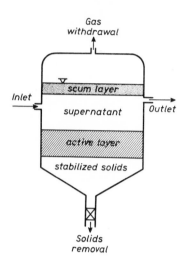

FIG. 6. Conventional anaerobic digestion system.

are increased, scum buildup becomes a problem, resulting in inferior supernatant, high in both BOD and suspended solids. Other problems and limitations become apparent when one begins to consider temperature control,
biological balance, and food contact. Because of the stratification, temperature control becomes a problem and true thermal homogeneity does not
exist. Because of stratification, microbial activity becomes restricted to
the scum and active sludge zones. Thus only a slender balance exists between the methane bacteria and the saprorophytic acid formers. Zoning also
limits contact between food and organisms.

C. High-Rate Units

High-rate digesters resulted from the efforts of many workers. Problems of temperature control, biological balance and food contact are avoided
by completely mixing the contents of the digestion tank (Fig. 7). This increases
the reaction zone from a fraction of the tank capacity to the full operating capacity. Mixing is achieved by either gas recirculation or mechanical agitation.

D. Anaerobic Contact Processes

The anaerobic contact unit developed out of the same demands as the high-
rate unit. A typical unit consists of an equalizing tank, completely mixed digester, and a settler. This work was pioneered by Fuller (1953), Steffen (1955)

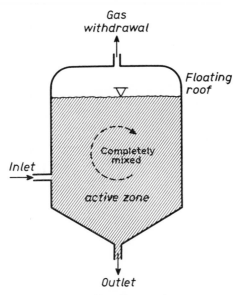

Gas
withdrawal

Floating
roof

Inlet

Completely
mixed

active zone

Outlet

FIG. 7. High-rate anaerobic digestion system.

and Schroepfer et al. (1955). In this process (Fig. 8) the high-rate system is modified by recycling of settled sludge from the reactor effluent back into the reactor feed line. This helps maintain a high microorganism population within the reactor, making it possible to achieve digestion with 24 hr or less contact. Generally, the recycle stream is 1 to 3 times the volume of the raw feed. In this process it is also necessary to have an equalizing tank before the reactor in order to avoid upsetting the ecological population of organisms. It is also necessary to degasify the reactor effluent to facilitate settling of the sludge (Schroepfer et al., 1955).

An anaerobic activated sludge process has been investigated with characteristics similar to aerobic activated sludge (Dague et al., 1966). This system included a completely mixed reaction tank, a settling tank, a sludge digestion tank, and provisions for sludge recycle to the completely mixed reaction vessel (Fig. 9). Design criteria have not been established. Anaerobic contact processes have been used in sewage and waste disposal (Coulter et al., 1957; Fall and Kraus, 1961). The kinetics of an anaerobic contact process have been developed by Gates et al. (1967). A material balance for this system must include recycle.

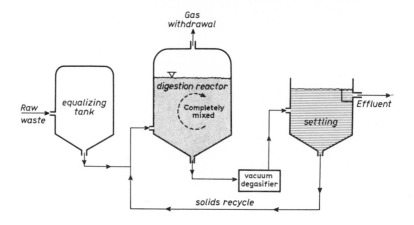

FIG. 8. Anaerobic contact process.

    The design features and operating characteristics of a conventional an-
aerobic process, a high-rate digester, and an anaerobic contact process are
compared in Table 7.

TABLE 7
DESIGN FEATURES OF ANAEROBIC DIGESTERS

	Conventional process	High–rate digesters	Anaerobic contact process
Heated or unheated	Heated, unheated	Heated	Heated
Detention time	>40 days	10-15 days	12-24 hr
Loading (lb VSS/ft^3/day)	0.03-0.05	0.1-0.2	0.1-0.2
Feeding & withdrawal	Intermittant	Continuous or intermittant	Continuous
Mixing	—^{a}	+	+
Feed equalization	—	—	+
Elluent sludge recycle	—	—	+
Degasification	—	—	+

^{a}Minus (—) not used; plus (+) used.

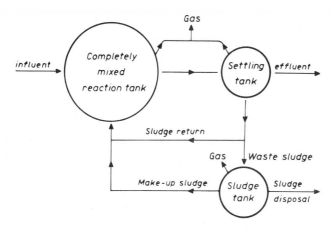

FIG. 9. Flow diagram for anaerobic activated sludge process.

IV. CURRENT DIGESTER PROBLEMS

Single reactor approaches which have been developed to date do not emphasize that two major ecological groups of bacteria are involved in bringing about anaerobic digestion. The generation time of the methane producers has been placed at several days. Since the growth rates of the methane bacteria are much slower than those of the volatile organic acid producing bacteria, they will control the feed and flow rates of a digester. Differences in pH optima, temperature optima, and competition of the two ecological groups for some similar nutritive compound indicates that organic acid production and methane synthesis should be carried out in two separate stages. This is also supported by Andrew and Pearson (1965), who developed constants for a two-stage kinetic model (Table 6).

Another critical problem encountered is startup. Careful analysis must be made of volatile organic acid buildup. Lime must be added any time the pH decreases to 6.5, or methane production is inhibited. Volatile solids loading also must be kept low--namely, 0.054 lb/ft^3/day for 40 days.

It may be that some of the older units have operated as two-stage processes where equalization tanks have been used to give the required acid formation and then coupled and fed into high microbial levels in the digester; thus all the apparatus is present for a two-stage process (Fullen, 1953; Steffen, 1955).

The pH of an anaerobic digester is related to several acid-base chemical reactions. Over a pH range of 6.0-8.0 the major controlling factors are carbon dioxide and bicarbonate ions. The hydrogen ion concentration is obtained from the following equilibrium equation:

$$\left[H\right] = K_1 \frac{\left[H_2CO_3\right]}{\left[HCO_3^-\right]}$$

The carbonic acid (H_2CO_3) is related to the CO_2 levels in the digestion, K is the ionization constant for carbonic acid, and the bicarbonate ion is related to the total alkalinity present in the system.

For most wastes the bicarbonate alkalinity is approximately equal to the total alkalinity, especially if volatile acids are low. As volatile acids increase they are neutralized by bicarbonate alkalinity and form what has been termed volatile acid alkalinity (Pohland and Bloodgood, 1963; McCarty, 1964):

BA = TA - (0.85) (0.833) TVA

Where BA = bicarbonate alkalinity, mg/liter as $CaCO_3$

      TA = total alkalinity, mg/liter as $CaCO_3$

    TVA = total volatile acids, mg/liter as acetic acid

  0.85 = 85% of volatile acid alkalinity measured by titration to pH 4.0

Lime is commonly added for control of pH in anaerobic digesters. It is cheap and does not affect anaerobic microbes adversely. At times, insoluble calcium salts form and cause some problems. In a single stage digester, problems occur if the pH decreases below 6.5. Liming increases pH by increasing bicarbonate alkalinity. Lime should not be added until the pH decreases to 6.5 and only enough should be added to bring the pJ to 6.7-6.8:

$$Ca(OH)_2 + 2CO_2 \longrightarrow Ca(HCO_3)_2$$

Calcium bicarbonate is not very soluble and when bicarbonate alkalinity reaches between 500-1000 mg/liter, the addition of lime results in the formation of insoluble calcium carbonate:

$$Ca(OH)_2 + CO_2 \longrightarrow CaCO_3 + H_2O$$

Liming at this point does not increase soluble bicarbonate alkalinity.

Temperature is important in digester operation. Rapid changes in operation should be completely avoided. Golueke (1958) found very little difference in digester operation at specified controlled temperatures between 35-60°C.

Factors examined were measured in terms of percentage destruction of volatile acid synthesis. Volatile acids increased significantly above 45°C, but
methane production did not decrease until temperatures of 60°C were exceeded.
No report was made at low temperatures, but unless the organism population
is adapted one would expect adverse affects in the process.

Any time the volatile acids exceed a concentration of 3000 ppm, it is indicative of overloading the digester, and methane production will decrease.
Any time adverse conditions affect the process, acid production increases,
indicating that the methane producing bacteria are far more sensitive then the
acid formers (Schulze and Raju, 1958).

Loadings of anaerobic digesters vary from 0.100-0.143 lb BOD/ft^3/day.
The gas produced is from 4.2-11.0 ft^3/lb. The volume of waste per tank
volume is 0.43-1.0 (Buswell, 1950).

Digester start up is critical. If no seed is available, the standard start
up procedure is as follows (Filbert, 1967):

1.   Fill the digester with sewage.

2.   Bring the temperature of the contents to the operating level.

3.   Start the mixing equipment.

4.   Close the system from the atmosphere. Do not light the waste gas burner
until all oxygen is flushed from the gas system. This may prevent a messy
explosion.

5.   Begin pumping sludge to the digester. The feed rate should be as uniform
as possible.

6.   Analyze digester contents for volatile acids, alkalinity, and pH, at least
daily. Make sure the samples analyzed are representative. Keep good records
of the above analyses, as well as temperature, gas production, and gas composition.

7.   Add alkalinity, the pH control agent, to the digester as required to keep
the total alkalinity concentration above the volatile acid concentration and the
pH above 6.6. Make every attempt to keep the pH near constant. Feed the
pH control agent and the raw sludge simultaneously. If possible, feed both
continuously.

REFERENCES

Andrews, J. F. and E. A. Pearson. 1965. "Kinetics and characteristics of volatile acid production in anaerobic fermentation processes." Int. J. Air and Water Pollution, 9, 439-461.

Barker, H. A. 1967. "Biochemical function of corrinoid compounds." Biochem. J. 105, 1.

Burbank, N. C. Jr., J. R. Cookson, J. Goeppner and D. Brooman. 1956. "Isolation and identification of anaerobic and facultative bacteria present in the digestion process." Int. J. Air and Water Pollution, 10, 327-342.

Buswell, A. M. 1950. "Operation of anaerobic fermentation plants." Ind. Eng. Chem. 42, 605-607.

Cantois, E. D. 1959. "Kinetics of bacterial growth - relationship between population density and specific growth rate of oontinuous cultures." J. Gen. Microbiol. 21, 40-50.

Carnegie Institute of Washington, 1953. "Algae culture from laboratory in pilot plant." Publ. 600, Carnegie Inst. of Wash., Wash. D. C.

Coulter, J. D., S. Soneda and M. G. Ettinger. 1958. "Anaerobic contact process for sewage disposal." J. Water Pollution Contr. Fed. 29, 468.

Dague, R. R., R. E. McKinney and J. T. Pfeffer, 1966. "Anaerobic activated sludge." J. Water Pollution Contr. Fed. 38, 220-226.

Fall, E. B. and L. S. Kraus. 1961. "The anaerobic contact process in practice." J. Water Pollution Contr. Fed. 33, 1038.

Filbert, J. W. 1967. "Procedures and problems of digester startup." J. Water. Pollution Contr. Fed. 39, 367-372.

Fullen, W. J. 1953. "Anaerobic digestion of packing waste." Sew. Ind. Waste 3, 576-585.

Gates, W. E., J. H. Smith, S. D. Lin and C. H. Ris, 3rd. 1967. "A rational model for the anaerobic contact process." J. Water Pollution Contr. Fed. 39, 1951-1970.

Golueke, C. G. 1958. "Temperature effects on anaerobic digestion of raw sewage sludge." Sew. Ind. Wastes 30, 1225-1232.

Herbert, D. 1961. "A theoretical analysis of continuous culture systems." Continuous Culture of Microorganisms. S.C.I. No 12, MacMillan, New York.

Herbert, D., R. Elsworth and R. C. Telling. 1956. "The continuous culture of bacteria: A theoretical and experimental." J. Gen. Microbiol, 14, 601-622.

James, T. W. 1961. "Continuous culture of microorganisms." Ann. Rev. Microbiol. 15, 27-46.

Jeris, J. S. and P. L. McCarty. 1962. "The biochemistry of methane fermentation using carbon-14 tracers." Proc. 17th Purdue Ind. Waste Conf. Purdue University; pp. 181-197.

Kester, A. S. and J. W. Foster. 1963. "Diterminal oxidation of long-chain alkanes by bacteria." J. Bacteriol 85, 859-869.

Langford, L. L. 1957. "Mesophilic anaerobic digestion, Part I. Fundamental factors of the process." Water and Sewage Works Oct., 464-469.

Leudeking, R. and E. L. Piret. 1959. "Transient and steady state in continuous fermentation: theory and experiment." J. Biochem. Microbiol. Technol. Eng. 1, 431-459.

Liubimov, V. I. and Z. S. Kagan. 1958. "The dynamics of volatile organic acids formed during anaerobic decomposition of organic compounds by microorganisms in methane tanks." Mikrobilogiya 24, 476-480.

Malek, I. 1961. "Development and further perspective of the continuous flow method cultivations." Continuous Culture of Microorganisms S.C.I. Monograph No. 12, MacMillan, New York.

Malek, I. and K. Beran. 1962. "Continuous cultivation of microorganisms." Folia Microbiol. 7, 388-411.

McCarty, P. L. 1964. "Anaerobic waste treatment fundmentals." Public Works Oct., 123-126.

McCarty, P. L. 1966. "Kinetics of waste assimilation in anaerobic treatment." Dev. Ind. Microbiol. 7, 144.

McCarty, P. L. and R. E. McKinney. 1961. "Volatile acid toxicity in anaerobic digestion." J. Water Pollution Contr. Fed. 33, 223-232.

Monod, J. 1950. "La technique de culture continue, theroie et applications." Ann. Inst. Pasteur 79, 390-410.

Moser, H. 1958. "The dynamics of bacterial populations maintained in the chemostat." Publ. 614, Carnegie Inst. of Wash., Wash. D. C.

Novick, A. and L. Szilard. 1950. "Experiments with the chemostat on spontaneous mutations of bacteria." Proc. Natl. Acad. Sci. (U.S.A.) 36, 708-719.

Pantskhava, E. S. and V. Y. Bykhovskii. 1965. "Biochemical and microbiological patterns of biosynthesis of vitamin B_{12} on thermophilic methane fermentation. Appl. Biochem. Microbiol. 1, 37.

Pfeffer, J. E., M. Leiter and J. R. Worland. 1967. "Population dynamics in anaerobic digestion." J. Water Pollution Contr. Fed. 39, 1305-1322.

Pohland, F. G. and D. E. Bloodgood. 1963. "Laboratory studies on mesophilic and thermophilic sludge digestion." J. Water Pollution Control. Fed. 1, 11-42.

Quinn, L. Y. 1962. "Continuous culture of ruminal microorganisms in chemical defined medium - I: Design of a continuous - culture apparatus." Appl. Microbiol. 10, 580-582.

Schroepfer, G. J., W. J. Fullen, A. S. Johnson, N. R. Ziemke and J. J. Anderson. 1955. "The anaerobic contact process as applied to packinghouse wastes." Sew. Ind. Wastes 27, 460-486.

Schroepfer, G. J. and N. R. Ziemke. 1958. "Development of the anaerobic contact process. Ancillary investigation and special experiments." Sew. Ind. Wastes 31, 697-711.

Schroepfer, G. J. and N. R. Ziemke. 1959. "Development of the anaerobic contact process. Pilot plant investigations and economics." Sew. Ind. Wastes 31, 2, 164.

Schulze, K. L. and B. N. Raju. 1958. "Studies on sludge digestion and methane fermentation. Methan fermentation of organic acids." Sew. Ind. Wastes 30, 164-184.

Steffen, A. J. 1955. "Full scale modified digestion of meat packing waste." Sew. Ind. Wastes 27, 1364-1368.

Toerien, D. F. 1967. "Enrichment culture studies on anaerobic and facultative anaerobic bacteria found in anaerobic digestion." Water Res. 1, 147-155.

Wood, W. A. 1961. "Fermentation of Carbohydrates." In The Bacteria, Ed. R. Stanier, Academic, New York.

Chapter 10

OXIDATION PONDS AND LAGOONS

Sewage ponds have been known in Asia for centuries. It is a well known
law of nature that the longer a contaminated body of water is held, the
greater the chance that natural oxidative, filtering bioconversions have of
returning the water to a fresh water condition. Data collected in Europe,
Asia, and the United States indicate that ponds provide a practical means of
treating sewage and certain wastes. Over 31 types of industrial wastes are
treated by waste stabilization pond processes (Porges, 1963 a,b). The conversion of sewage to edible alga protein is feasible (Gotaas et al., 1954) and
is receiving considerable developmental study.

Oxidation ponds are large and shallow with wastes being added at a single
point at about one-third the length of the pond. The effluent is removed at a
single point at the end of the pond. Organic wastes are best treated in this

manner. The purification is a natural process catalyzed by biological agents, i. e. it takes advantage of the microbes and algae that are present that catalyze the oxidation and removal of the organics. Lengthy storage of the waste water in a controlled environment is required.

I. THEORY OF OPERATION

The removal of organic matter from oxidation ponds is brought about through the metabolism of two major groups of microbes: hetrotrophic microbes, which oxidize organic materials for energy, and photosynthetic algae, which fix CO_2 for cellular carbon and derive their energy from the sunlight (Zajic, 1970). Recent studies on the algae indicate that many of the more common soil and water forms also may grow heterotrophically as well as photosynthetically or by both modes of metabolisms (mixotrophically). Normally, heterotrophic microbes and algae are not regarded as being competitive for the same nutritive materials. Under normal conditions, they compliment each other because the algae produce oxygen as an end product, which benefits the growth of aerobes and facultative anaerobes (Fig. 1). The oxygen is used by the microbes as a terminal electron acceptor for oxidizing organic wastes to CO_2 and water (n=1 in this example):

$$(CH_2O)_n + O_2 \longrightarrow CO_2 + H_2O$$

waste

The microbes produce additional cellular material and enzymes that catalyze the oxidation of the organics. The CO_2 produced is in turn used by the algae.

The algae do not produce adequate oxygen for the heterotrophic population of microbes, and supplementary oxygen must be obtained by surface aeration of the pond. Atmospheric reaeration is approximated by the empirical equation of Imhoff and Fair (1940).

$$R = 0.0271 \, adD_o \tag{1}$$

where R is reaeration, lb/acre/day

       d is pond depth, ft

       D_o is the 24-hr average dissolved oxygen deficit

       a is an arbitrary factor, 20

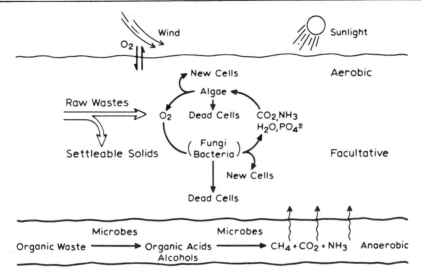

FIG. 1. Schematic diagram of oxidation pond operation.

Common algae in waste water are reviewed in detail by Palmer (1962).
A few common forms are presented in Fig. 2. In the Chrysophyta the entire
aggregation of mobile unicellular and colonial forms are represented. This
includes Synura, Botryococcus, and Mallomonas, and the planktonic diatoms
such as Asterionella, Cyclotella, Fragellaria, and Synedra. Members of the
Cyanophyta, notorious as nuisance microbes, are also present. These are
Diplocystis (Microcystis), Aphanizomenon anabaena, Lyngbya, and Oscilla-
toria.

The overall equation for the growth of algae is expressed for Chlorella by

$$NH_3 + 5.7\ CO_2 + 12.5\ H_2O \xrightarrow{\lambda} C_{5.7}H_{9.8}O_{2.3}N + 6.25\ O_2 + 9.1\ H_2O \qquad (2)$$

Typical sewage and pond algae are shown in Fig. 2(a), (b). The microbial
populations in oxidations ponds are far too complex for simple description,
although in time even this complexity must be studied in terms of ecological
groups of microbes. Pseudomonas, Flavobacterium and Alcaligenes are
representative genera of the bacteria. Of the fungi, Penicillum, Aspergillus,
Fusarium, and many other genera are present (Cooke, 1963). Pathogenic
microbes normally decrease in lagooning operations. Coliform reductions
of up to 98% are common (Mackenthum and McNabb, 1961).

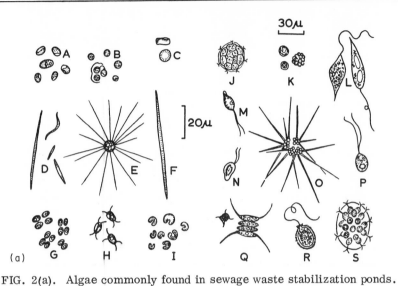

FIG. 2(a). Algae commonly found in sewage waste stabilization ponds.
A, Chlorella ellipsoidea; B, Chlorella vulgaris; C, Cyclotella operculata;
D, Ankistrodesmus spp.; E, Golenkinia radiata; F, Nitzschia palea; G, Cru-
cigenia rectanularis; H, Chodatella quadriseta; I, Selanastrum minutum;
J, Pandorina morum; K, Coelastrum microporum; L, Euglena spp.; M, Chlor-
ogonium euchlorum; N, Cryptomonas erosa; O, Micractinium pusillum,
P, Chlamydomonas sp.; Q, Scenedesmus spp.; R, Trachelomonas sp.;
S, Eudorina elegans.

II. POND CLASSIFICATION

Oxidation ponds are classified by the nature of their biological activity
into aerobic, facultative, and anaerobic ponds. Even within an aerobic
pond, zones exist where faculative or anaerobic conditions predominate
(Fig. 1). In aerobic ponds, dissolved oxygen should always be present in at
least a concentration of 2 ppm. As dissolved oxygen becomes limiting, facultative anaerobic conditions become established. Facultative anaerobes prefer
anaerobic conditions but do utilize and compete for small amounts of dissolved
oxygen.

Facultative anaerobes and anaerobes use compounds other than gaseous
O_2 as terminal electron acceptors. Fortunately the oxygen present in a
variety of compounds can be used in place of oxygen to carry out oxidative
reactions. Compounds acting in this capacity are almost any oxidized organic

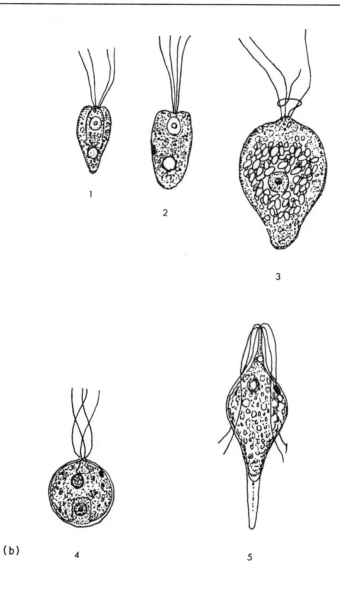

FIG. 2(b). Examples of sewage and pond algae: 1, Pyraminonas prox. tetra-hynchus; 2, Pyraminonas prox. montana; 3, Polytomella citri; 4, Carteria sp. ; 5, Pyrobotrys gracilis (lateral view of full grown zygote without arms).

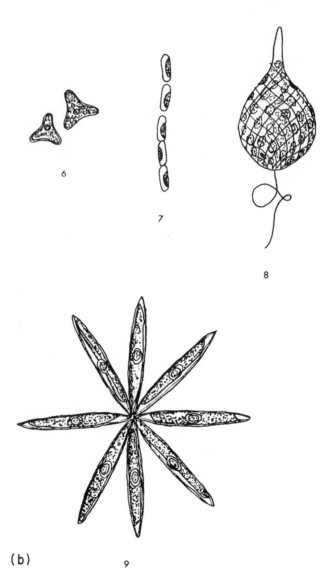

(b)

FIG. 2(b) cont'd. 6, Tetraedron muticum; 7, Ulotrichales: Stichococcus bacillaris; 8, Phacus sp.; and 9, Actinastrum hantzschii var fluviatile.

molecule or inorganic molecular such as nitrate and sulfate. In a typical anaerobic pond no dissolved oxygen is present. Anaerobic conditions are normally avoided because the anaerobic microbes produce a large number of little-known odoriferous compounds. The depths of aerobic ponds, the most common ponds used, vary from 3 - 6 ft.

Aerobic ponds are in essence the upper layer of a facultative pond. In theory the depth of the pond should be limited to the level that light penetrates, particularly if algae are desired. For most wastes this does not exceed 18 in. High concentrations of algae are maintained in these ponds. The algae concentration should be high enough to prevent light penetration to the bottom of the pond. This prevents growth of undesirable weeds. To maintain aerobic conditions and to resuspend any settled sludge, some type of mixing is often practiced and operated 4 - 8 hr per day.

The organic loading of anaerobic ponds is so high that dissolved oxygen is depleted to such an extent that anaerobic conditions exist throughout the pond. The biological processes are the same as those occurring in anaerobic digestion tanks, i.e., primarily organic acid formation followed by methane syntheses. The depth of anaerobic ponds is selected to give a minimum surface-area-to-volume ratio and thereby provide maximum heat retention.

A new type of pond introduced prior to 1960 is the "aerated lagoon." It developed because of a need for a treatment process with the operational characteristics of the oxidation pond, but with the ability to treat more organic matter per unit volume. Oxidation ponds giving erratic performance characistics may be improved by utilizing forced aeration.

An aerated lagoon is a basin of significant depth (6-12 ft) in which oxygenation is accomplished by mechanical or diffused aeration units and by induced surface aeration. The turbulence level maintained in the pond ensures distribution of oxygen throughout the pond. The microbial characteristics of the aerated lagoon are those of an activated sludge system rather than an oxidation pond. The turbidity and turbulence prevent normal alga growth except in the upper levels.

Since the active biological solids level in an aerated lagoon is low, BOD removal is primarily a function of retention time, temperature, and the nature of the waste. Unless controlled channelling is built into the lagoon, lagoons

are treated as approaching a completely mixed regime. The removal of BOD
can usually be formulated according to first order kinetics:

$$\frac{Le}{Lo} = \frac{1}{1 + Kt} \qquad\qquad (3)$$

where   Lo is the initial 5-day BOD, ppm

Le is the final 5-day BOD, ppm

t is the retention time, days

K is BOD removal rate coefficient

K depends upon temperature in accordance with the equation

$$K_t = K_{20^{\circ}C} \; \theta^{(T - 20)} \qquad\qquad (4)$$

The coefficient θ for domestic sewage is about 1.035.

Microorganisms in aerated lagoons metabolize most of the organics in the
raw sewage with the production of an appreciable quantity of cellular material.
The continuous addition of raw sewage results in the displacement of mixed
liquor containing microbial solids. The major source of BOD in the effluent
from an aerated lagoon is related to this displacement of active microbial
solids. The key to the effluent quality in the aerated lagoon lies in the removal
of these solids. It is therefore prudent at times to couple an aerated lagoon
with an oxidation pond. The excess cells will be discharged in the effluent
from the aerated lagoon to the oxidation pond where they will undergo sedimentation and eventual lysis. Cells in this sediment enter into the declining phase
of growth where endogenous substrates are reduced to a minimum.

III. DESIGN CRITERIA

An idealized projection of a stabilization pond is shown in Fig. 3
(Oswald, 1961-63). Most oxidation ponds are designed on the basis of pounds
of BOD loading per acre per day, depth, and detention time. Engineering
experience and the cost of land are the primary controlling factors. Fundamental factors involved in waste stabilization design are reviewed by Oswald
(1961-63) and Caldwell (1946). A summary of design factors for the various
types of ponds is presented in Table 1.

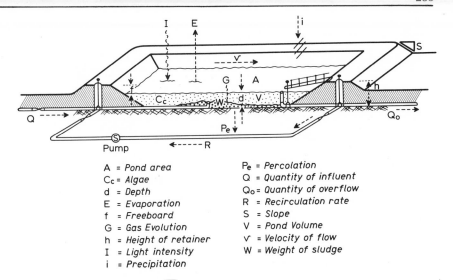

A = *Pond area*
Cc = *Algae*
d = *Depth*
E = *Evaporation*
f = *Freeboard*
G = *Gas Evolution*
h = *Height of retainer*
I = *Light intensity*
i = *Precipitation*

Pe = *Percolation*
Q = *Quantity of influent*
Qo = *Quantity of overflow*
R = *Recirculation rate*
S = *Slope*
V = *Pond Volume*
v = *Velocity of flow*
W = *Weight of sludge*

FIG. 3. Idealized projection of a stabilization pond showing design variables and symbols (after Oswald, 1961-63).

TABLE 1

OXIDATION POND DESIGN CRITERIA

Factor	Aerobic	Facultative	Anaerobic	Aerated
Depth (ft)	0.6-1.0	2-5	8-10	6-15
Detention (days)	2-6	7-30	30-50	2-10
BOD loading (lb/acre/day)	100-200	20-50	300-500	--
BOD removal (5)	80-95	75-90	50-70	55-80
Algae concentration (ppm)	> 100	10-50	nil	nil

Oswald and Gotaas (1955) have applied the following equation to the design of aerobic ponds:

$$A = \frac{hWO_2}{pES} \tag{5}$$

where   A is the surface area of pond, cm^2

h is the unit heat of combustion, cal/gm

WO_2 is the net weight of oxygen produced, g/day

p is the oxygenation factor

E is the efficiency of energy conversion

S is the solar radiation in Langleys, cal/cm^2 per day

Examples 1 and 2 show how to determine the variables in this basic equation and solve it for surface area. It should be noted that these ponds are applicable where optimum light intensities prevail.

Hermann and Gloyna (1958) devoted several years to experimental and statistical studies of oxidation ponds in Texas. Their studies resulted in a formula for determining the volume of facultative ponds:

$$V = 5.37 \times 10^{-8} \, Nqy(1.072^{35-T}) \tag{6}$$

where   V is the pond volume, acre/ft

N is the number of people served

q is the flow in gal/day/capita

y is the 5 day $20^\circ C$ BOD, mg/liter

T is the operating temperature, $^\circ C$

Climatic conditions limit application of this formula and it is not recommended that it be used for cold climates. Example 1 presents the design of a facultative pond. Anaerobic ponds are usually coupled with facultative ponds. A detention time of 3-5 days in the anaerobic pond will remove 50-70% of the BOD. This will substantially reduce the BOD load and acreage of the facultative pond, which will serve for polishing the effluent. An anaerobic pond should never be used unless the waste is amenable to treatment by anaerobic processes.

Aerated lagoons are designed according to first order kinetics as formulated in Eq. (3). The design procedure is presented in Example 2.

The most common aeration system applied to these lagoons is the mechanical surface aerator. The effective range of each aerator is a radius of about 35-50 ft. Diffused aerators of the type used in the activated sludge processes are generally not applicable to aerated lagoons. For effective results this system requires that air headers be spaced at a distance of twice the water depth, and the spargers at a spacing of 12-18 in. Normally the air demand requirements of an aerated lagoon require a much larger spacing of headers.

A unique air diffusion technique, developed by Hinde Engineering Company, and commercially known as the "AIR-AQUA' oxidation system, is commonly utilized in diffused aeration. The system is usually applied in complete treatment processes through progressive multistage aeration. The system consists of a blower, PVC manifold, and specially formed and processed polyethylene aeration tubing with die-formed check valves for the release of controlled pin-point air bubbles (Figs. 4 and 5).

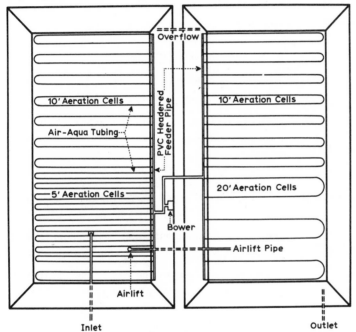

FIG. 4. Typical air-aqua system (plan view).

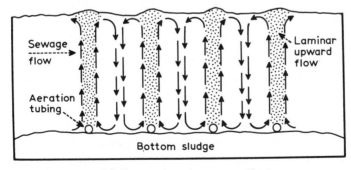

FIG. 5. Air-aqua oxidation system (cross section).

IV. OPERATIONAL ASPECTS

A. Level Control

The ability to alter the level of an oxidation pond is vital in pond treatment. Through level control, storage operation can be achieved, especially during the winter months when biological activity is reduced. Flexible operations of series or parallel treatment flow in multiple ponds should be arranged as desired. In attaining level control, the simplest type of mechanical devices are preferred. The stop-log weirs are the most common.

B. Berm Maintenance

Erosion control appears to be the main requirement of berm maintenance. Both wave and runoff erosion can be effectively diminished through the development of a vigorous grass cover and by utilizing a gentle grade on berm slopes. Slopes of 4:1 are recommended.

C. Odors

Odor problems are rarely encountered in stabilization ponds unless the BOD exceeds 20 lb/acre/day. Some industrial wastes have a tendency to produce odors. In colder climates the springtime ice breakup is generally the most critical period of the year for odors. Severe Ice Breakups last about two weeks. The application of sodium nitrate or the addition of aerated river water just prior to the odor period is beneficial.

D. Weeds and Insects

Both aquatic and terrestial weeds are encountered in oxidation ponds. If maintenance in control of weeds is not efficient, the ponds will develop into mosquito breeding areas. Several commercial herbicides are available for spraying aquatic weeds. In general, triazine-type herbicides have proved most successful in the control of aquatic weeds. The terrestial weeds can be controlled by mowing and the use of selective herbicides.

Of the weeds, Tule (Sciptus spp.) and cat-tail (Typha spp.) cause the most problems in ponds. The plants do not take root below 24 in. If mosquitos become a nuisance, weeds must be removed along the shore line and the fish feeding on mosquito larvae such as Gambusia affinis may be introduced into the pond.

E. Series and Parallel Operation

The question of arranging ponds in parallel or series ponds has not been resolved. The purpose of the series ponds is to effect better circulation of the wastes into the pond and to give the greatest total retention time; there is no reported advantage to series ponds over parallel ponds. Certain advantages of series ponds are claimed when the organic load is raised considerably above design criteria.

Laboratory model studies indicate that the assumption that most ponds operate as completely mixed systems is incorrect (Thirumurthi, 1969). Rectangular ponds approach plugflow. The flow patterns are of a dispersed plugflow category and they are more efficient than square ponds. If ponds are used solely as percolation and evaporation beds, cost is minimized by reducing the number of ponds. If flow patterns are required and essential to foster various biological activities, multiple ponds have advantages. Large ponds of 20-acre capacity always require greater levee maintenance. There is increased sepage, greater freeboard is required, and thus the height of the levee and costs are increased.

F. Recirculation

Recirculation of pond effluent will add oxygen and algal seed at the influent end of the pond. Recirculation from 0.2 to 1.5 times the pond volume, depending on the strength of the sewage, is desirable.

G. Autoflocculation

This occurs in ponds under special environmental conditions. Usually, dense algal populations are observed, following an increase in the temperature and a rise in the pH of the water. If adequate magnesium and calcium ions are present, salts of magnesium hydroxide and ammonium calcium phosphate precipitate out. Algal cells, bacteria, and organic debris are enmeshed within the precipitate to form larger particles, which settle rapidly. Much of the hardness of the water is removed, and since algae are carried out of the photosynthetic zone, a temporary decrease in dissolved oxygen is observed. The sludge that settles continues to undergo biodegradation.

The average composition of gases evolved from bottom sludge from a facultative pond is 39% CH_4, 1.0% O_2, 2.0% CO_2, 12.5% H_2, and 45.5%

unidentified (Oswald, 1961-1963). Gases and organic by products are formed during biodegradation.

H. Toxins

Toxin production from algae can be a serious problem. In those areas where known toxic algae are reported, the practice of permitting livestock to water on algae ponds should be avoided (Schantz, 1970).

I. Suppression of Photosynthetic Oxygenation

The presence of excessive concentrations of toxic organics that appear as a result of periodic slug releases of effluent waste may inhibit the production of photosynthetic oxygen and thereby possibly decrease the effectiveness of the oxidation pond.

A long-range study to determine the relative toxicity of selected organic compounds revealed the following information. Straight chain fatty acids with an odd number of carbon atoms seem to be more toxic than the corresponding acids with even number of carbon atoms. Straight chain fatty acids with four and five carbon atoms seem to be more toxic than the corresponding branched chain acids. Toxic effects of dicarboxylic acids (C_2 to C_7) appear to decrease with increasing number of carbon atoms per molecule for the first three acids (C_2, C_3, C_4) in the chain. The toxicity increases with increasing number of carbon atoms for the fourth, fifth, and sixth members. Octanol at a concentration of 250 mg/liter exerts a toxic effect, whereas methanol is not toxic until a concentration of 32,500 mg/liter is reached. The toxicity of alcohols to the algal cell increases with increasing molecular weights. Many compounds, e.g. aldehydes, bisulfate, phenol, are toxic to microbes and if present they must be handled with care to avoid shock loading (Ravenna, 1962).

J. Costs

Capital costs of oxidation ponds range from $3000 to $6000 per acre. Oxidation ponds now operating in Ontario have cost from $14 to $58 per capita at the time of construction.

Example 1: Calculations for a Facultative Pond and for a
Mechanical-Aerator Type Aerated Pond.

1. **Data**

   Flow $= 0.6$ I.M.G.D.

   BOD_5 $= 200$ ppm

   BOD_5 Load $= 0.6 \times 10 \times 200 = 1200$ lb

2. **Surface Area of Facultative Pond**

   At a loading of 20 lb BOD_5/acre/day, surface area of pond $= 1200/20 =$ 60 acres. For practical reasons, 4 ponds of 15 acres each are chosen. By coupling an aerated pond with a facultative pond (for the same example), the former will render an intermediate treatment, and the latter the final polishing.

3. **Volume of Proposed Aerated Pond**

   Assume the removal efficiency approximates first-order kinetics and the desirable retention time is 4 days:

   $$\frac{Le}{Lo} = \frac{1}{1 + Kt}$$

   Let $K_{15°C} = 0.35$. Then $\frac{Le}{200} = \frac{1}{1 + 0.35 \times 4}$, Le $= 83$ ppm and

   $E = \frac{(200 - 83)}{200} \times 100 = 60\%$. Volume of proposed aerated pond: 4 days x 0.6

   I.M.G.D. $= 2.4$ I.M.G. $= 385,000$ ft^3.

4. **Pond Geometry**

   Depth $= 11$ ft

   Surface Area $= 220$ ft x 220 ft ($= 1.1$ acres)

   Bottom Area $= 154$ ft x 154 ft (Berm slope of 3:1)

5. **Oxygen Demand**

   From practice, oxygen requirement for BOD_5 removal ranges between 1.1 and 1.5 lb/lb of BOD_5 removed. Accepting the higher value, the oxygen demand will be: $(0.6 \times 10 \times 200) \times 0.6 \times 1.5 = 1080$ lb/day

   $$= 45 \text{ lb/hr}$$

6.   Mechanical Aerators Proposed

Oxygen transfer rate of mechanical aerators ranges from 2-2.5 lb/hr/ hp. Total hp required:

$$\frac{45}{2} = 22.5 \text{ hp}$$

$$\frac{45}{2.5} = 18 \text{ hp}$$

Use 4 mechanical aerators of 5 hp each. Note that surface aeration was neglected. This serves as a safety margin.

7.   Surface Area Required for Polishing Pond

Unremoved BOD_5 load:

(0.6 x 10 x 200) x 0.4 = 480 lb

At a loading of 20 lb/acre/day, the required surface area would be

$$\frac{480}{20} = 24 \text{ acres}$$

Thus, coupling of an aerated lagoon with a facultative pond requires a total of 24 acres, compared with the 60 acres required for a conventional facultative pond.

Example 2: Method for Designing Aerobic Ponds

$$A = \frac{hWO_2}{PES}$$

where   A is the surface area of pond, cm^2

h is the unit heat of combustion, cal/gm

WO_2 is the net weight of oxygen produced, g/day

p is the oxygenation factor

E is the efficiency of energy conversion

S is the solar radiation in Langleys, cal/cm^2 per day

Determination of the variables in this basic equation is given here.

1. Unit heat of combustion, h

This is a function of the composition of the organic matter constituting the algal cell. It has been found to be approximately 6000 cal/g (on an ash-free basis). To compute the value of h, the following empirical expression is used:

$$h = 127R + 400$$

where R is an expression of the chemical composition of the algae:

$$R = \frac{100 \left[(2.66 \times \% \text{ Carbon}) + 7.94 \ (\% \text{ Hydrogen}) - (\% \text{ Oxygen})\right]}{393.9}$$

(The percentages of carbon, hydrogen, and oxygen should be computed on an ash-free weight basis.)

2. Oxygenation factor, p

This is the ratio of the weight of oxygen produced to the weight of algal cells synthesized. It will therefore vary with the composition of the algal cells. Under most environmental conditions, the ratio has been found to range in value. The chemical composition of the algal cells, must be determined as well as the oxygen produced:

$$aCO_2 + (0.5b-1.5d)H_2O + dNH_3 \longrightarrow C_aH_bO_cN_d + (a+0.25b-0.75d-0.5c)O_2$$

The value of p is

$$\frac{(a + 0.25b - 0.75d - 0.5c)\ O_2}{C_aH_bO_cN_d}$$

The subscripts a,b,c,d are known from the analysis of the chemical composition of the algal cell. They must be multiplied by the atomic weight of the respective element.

3. Solar radiation, S

When incident on a horizontal surface, S depends upon the geographical location, the elevation, the season, and the meteorological conditions. Energy values for the visible portion of the spectrum (4000 - 7000Å) to be expected at various latitudes in the northern hemisphere during the year are presented in Table 2.

TABLE 2

SOLAR RADIATION IN LANGLEYS (cal/cm^2/day)[a]

Month

Latitude		Jan.	Feb.	Mar.	Apr.	May	Jun.	Jul.	Aug.	Sep.	Oct.	Nov.	Dec.
0	max	255	266	271	266	249	236	235	252	269	265	256	253
	min	210	219	206	188	182	103	137	167	207	203	202	195
10	max	223	244	264	271	270	262	265	266	266	245	228	225
	min	179	184	193	183	192	129	153	176	196	181	176	162
20	max	183	213	246	271	284	284	282	272	252	224	190	182
	min	134	140	168	170	194	148	172	177	176	150	138	120
30	max	136	176	218	261	290	296	289	271	231	192	148	126
	min	76	96	134	151	184	163	178	166	147	113	90	70
40	max	80	130	181	181	286	298	288	258	203	152	95	66
	min	30	53	95	125	162	173	172	147	112	72	42	24
50	max	28	70	141	210	271	297	280	236	166	100	40	26
	min	10	19	58	97	144	176	155	125	73	40	15	7
60	max	7	32	107	176	249	294	268	205	126	43	10	5
	min	2	4	33	79	132	174	144	100	38	26	3	1

[a]after Oswald, 1961–63.

The figures in Table 2 have to be corrected for cloudiness and elevation. The correction for cloudiness is

$$S_c = S_{min} + r(S_{max} - S_{min})$$

where $r = \dfrac{\text{total hours of sunshine}}{\text{total possible hours sunshine}}$

The correction for elevation up to 10,000 ft is

$$S_e = S_c (1 + 0.01 e)$$

where e is the elevation in feet.

4.   Fraction of energy available for photosynthesis

In visible light (Oswald and Gotaas, 1955) this is

$$f = \frac{I_s}{I_o} \left[\ln \left(\frac{I_o}{I_s} \right) + 1 \right]$$

where   f is the fraction of available light utilized

I$_o$ is the incident light intensity

I$_s$ is the saturation intensity

Energy conversion efficiencies may be limited by nutritional deficiencies of the effluent and adverse environmental conditions. Values of f may vary from 0.02 to 0.09, with an average value of 0.04 (Oswald, 1957).

5.   Rate at which oxygen WO$_2$ is produced

This must balance the rate at which oxygen is utilized in the oxidation of the organic wastes. The oxidation rate is a function of the time the organics are present in the system. A decision must be made as to the degree of treatment desired. This degree of treatment establishes the required retention time, which in turn establishes the rate at which oxygen must be supplied. The rate at which oxygen is required to support the aerobic decomposition taking place in an oxidation pond is estimated from the relationship commonly used for expressing the rate at which oxygen is utilized in the standard BOD test:

$$y_t = L (1 - 10^{-kt})$$

where y_t is the oxygen demand exerted in t days

L is the first stage or ultimate oxygen demand

k is the deoxygenation constant

t is the time, (days)

The deoxygenation constant k will vary depending upon the type of waste and temperature. For domestic sewage at $20°C$, k is assumed to be equal to 0.1 day^{-1}. The effect of temperature on the constant is approximated by

$$k_T = k_{20°C} \, (1.047^{T-20})$$

where T is the temperature in $°C$. Initial BOD removal is also influenced by temperature. The temperature effect has been formulated as

$$L_T = L_{20°C} \left[1 + 0.02(T-20) \right]$$

6.   Computation of the depth d of the pond

For this, we use the nominal relationship of retention time to pond capacity:

$$t = \frac{Ad}{Q}$$

For example,

Design criteria for oxidation pond:

Flow, 0.5×10^6 gal/day

$BOD_{5\ 20°C}$, 200 mg/liter

Locality, $40°N$ latitude

Elevation, 100 ft

Critical month, October, with expected temperature of $20°C$

$r = \dfrac{\text{total hours of sunlight}}{\text{total possible hours of sunlight}}$ 0.60

Photosynthesis energy conversion efficiency 0.04

REFERENCES

Caldwell, D. H. 1946. "Sewage oxidation ponds--performance operation and design." Sewage Works. J. 18, 433-458.

Cooke, Wm. B. 1963. A Laboratory Guide to Fungi in Polluted Waters, Sewage, and Sewage Treatment Systems. U. S. Dept. Health Ed. Welfare, PHS No. 999 - WP - 1.

Gotaas, H. B., W. J. Oswald and H. F. Ludwig. 1954. "Photosynthetic reclamation of organic wastes." Sci. Monthly 79, 368-378.

Hermann, E. R. and E. F. Gloyna. 1958. "Waste stabilization ponds III. Formulation of design equations." Sewage Ind. Wastes 30, 963-975.

Imhoff, K., and G. M. Fair. 1940. Sewage Treatment. Wiley, New York.

Mackenthum, K. N. and C. D. McNabb. 1961. "Stabilization pond studies in Wisconsin." J. Water Pollution Control Fed. 33, 1234-1251.

Oswald, W. J. 1957. "Algae in waste treatment." Sewage Ind. Wastes 29, 437-455.

Oswald, W. J. 1961-63. "Fundamental factors in stabilization pond design." Air and Water Pollution 5, 357-393.

Oswald, W. V. and H. B. Gotaas. 1955. "Photosynthesis in sewage treatment." A.S.C.E. Proc. 81, Sept. No. 686.

Palmer, C. M. 1962. Algae in Water Supplies. U. S. Dept. Health Ed. Welfare PHS No. 657.

Porges, R. 1963a. "Industrial waste stabilization ponds in the U. S." U. S. Publ. Health Service Paper. J. Water Pollution Control Fed. 35, 456-468.

Porges, R. 1963b. "Waste stabilization ponds: Use, function, and biota." Biotechnol. Bioeng. 5, 255-273.

Ravenna, V. 1962. "The effect of 17 toxicants on the growth of five species of algae." Appl. Microbiol. 10, 532.

Schantz, E. J. 1970. "Biochemical studies in algae toxins." In Properties and Products of Algae. Ed. J. E. Zajic, Plenum, New York.

Thirumurthi, D. 1969. "Waste stabilization ponds: Shape vs performance." Eng. Digest 15, 31-34.

Wiedeman, V. E. 1970. "Heterotrophic growth of waste-stabilization pond algae." In <u>Properties and Products of Algae</u>. Ed. J. E. Zajic, Plenum, New York.

Zajic, J. E. (Ed.) 1970. <u>Properties and Products of Algae</u>. Plenum, New York.

Chapter 11

SOLID WASTE DISPOSAL AND COMPOSTING

The storage, collection, and disposal of solid wastes is a major problem
in an urban area. In annual costs, that appropriated for sanitation and waste
disposal ranks next to education, debt charges, and police and fire protection.
Increased growth of the population and the continued move from the farm to

urban communities are putting a dangerous strain on solid waste disposal.
Improper methods of operation are a definite health hazard.

Solid wastes can roughly be divided into salvable material, waste for
sanitary landfills, and garbage. In developing information concerning solid
wastes, the study area should first be defined. Population data are needed
and the area must be defined in terms of industry, agricultural potential,
major markets, and potential for future growth. These items lead to long-
range planning as facilities and services are required for water supply,
sewerage, refuse disposal, flood control, transportation, highways, schools,
shopping centers, parks, industrial plants, homes and apartments. Services
for all must be planned for orderly growth or for zero growth.

In any comprehensive engineering or planning study the needs of the community must be projected for 20-25 years. Census statistics are necessary.
A precise method of making population projections must be selected. One
used by the Bureau of Statistics is the "cohort-survival method." Many areas
are increasing at a uniform geometric or compound rate.

Solid wastes must be classified. Many municipalities collect garbage
and refuse separately. Garbage refers to household waste and is usually animal or vegetable waste resulting from the handling, preparation, cooking,
and serving of food. It consists largely of putrescible organic matter. Refuse
refers to brush and garden trimmings, lawn cutting, Christmas trees, and
leaves. These terms are often used interchangeably. The American Public
Works Association (APWA) has standardized these terms. Waste means useless, unwanted, or discarded materials from normal community activities.
Wastes may be solid, liquid, or gas. Refuse comprises all solid waste of the
community. It includes semiliquid or wet wastes with insufficient moisture
or liquid to be flowing. The classification of waste as defined by APWA is
summarized in Table 1.

Collection of waste will not be discussed in any detail here. However, it
should be emphasized that the greater part of the cost of waste management
is in the collection of refuse, which may be 80% of the total. Thus, new methods of collection may offer the greatest opportunity for economy. Simulation
analyses of refuse collection systems are reviewed by Quon et al. (1965).

TABLE 1

REFUSE MATERIALS BY KIND, COMPOSITION, AND SOURCES

Kind		Composition	Sources
Refuse	Garbage	Wastes from preparation, cooking, and serving of food; market wastes; wastes from handling, storage, and sale of produce	
	Rubbish	Combustible: paper cartons; boxes, barrels, wood, excelsior, tree branches, yard trimmings, wood furniture, bedding, dunnage Noncombustible: metals, tin cans, metal furniture, dirt, glass, crockery, minerals	Households, restaurants, institutions, stores, markets
	Ashes	Residue from fires used for cooking and heating and from on-site incineration	
	Street refuse	Sweepings, dirt, leaves, catch basin dirt, contents of litter receptacles	
	Dead animals	Cats, dogs, horses, cows	Streets, sidewalks, alleys, vacant lots
	Abandoned vehicles	Unwanted cars and trucks left on public property	
	Industrial wastes	Food processing wastes, boiler house cinders, lumber scraps, metal scraps, shavings	Factories, power plants
	Demolition wastes	Lumber, pipes, brick, masonry, and other construction materials from razed buildings and other structures	Demolition sites to be used for new buildings, renewal projects, expressways
	Construction wastes	Scrap lumber, pipe, other construction materials	New construction, remodeling
	Special wastes	Hazardous solids and liquids: explosives, pathological wastes, radioactive materials	Households, hotels, hospitals, institutions, stores, industry
	Sewage treatment residue	Solids from coarse screening and from grit chambers; septic tank sludge	Sewage treatment plants; septic tanks

I. DISPOSAL METHODS

Although processing and disposal only account for 20% of the cost, the problems offered are more critical. In analyzing disposal methods the following must be considered: technical feasibility, limitations, effect of local conditions, public health, potential nuisance, weather, collection procedure, locations available, costs, flexibility, possible change in type of waste, and public opinion. Methods used are:

1. Open dumping and burning.
2. Sanitary landfill.
3. Central incineration.
4. On-site incineration.
5. Garbage grinding.
6. Feeding garbage to swine.
7. Composting.
8. Salvage and reclamation.

Open dumping and burning is no longer permitted. State legislaures have set controlling regulations. Waste management usually falls under the control of the Public Health Engineering Service of the Department of Health.

Landfill sites are closely regulated, particularly as to what cannot be used. In typical sanitary landfills the following items are prohibited:

1. Explosives or highly combustible material.
2. Car bodies.
3. Sheet iron and other scrap metal.
4. Tree stumps.
5. Corrosive or toxic materials.
6. Any material constituting a hazard to safety of personnel or damage to equipment.
7. Carcasses of animals larger than a dog.
8. Waste building materials unless specifically permitted.

It is estimated that total refuse production in the United States will have increased from 150 million tons a year in 1963 to 260 tons in 1980. Figure 1 shows the increase in refuse projection to 1980. The quantity of refuse is predicted to increase from 4.4 lbs/capita/day in 1963 to 5.8 lbs/capita/day in 1985 (lower part of Fig. 1).

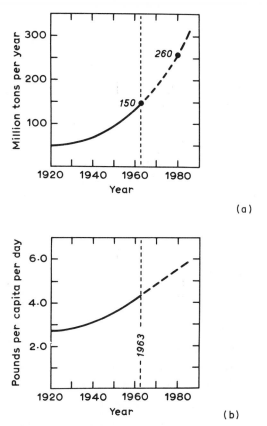

(a)

(b)

FIG. 1. (a) Total refuse production in the United States, (b) per capita refuse production in the United States. Prepared as part of a report to the Surgeon General's Advisory Committee on Urban Health Affairs: Solid Wastes Handling in Metropolitan Areas, Division of Environmental Engineering and Food Protection, Public Health Service, Department of Health, Education and Welfare, February 1964, 41 P.

Solid wastes from cities are often categorized as (a) municipally collected, (b) privately collected, (c) solid burning waste, (d) building rubble, (e) waste timber and trees, (f) ash from power generating stations, (g) street sweepings, and (h) manure. Domestic waste is a totally heterogenous material and will remain so unless segregation practices are required and enforced. The U.S. Public Health Service divides garbage into eight major categories: paper, food wastes, metal, glass, bulk leaves and grass, wood, plastic, and cloth (rubber and leather). Use of these divisions will permit

comparison between cities. A U.S. Department of Health Study made of the
quad-city area near Cincinatti summarizes the solid waste composition from
four cities (Table 2).

TABLE 2

QUAD-CITY MUNICIPAL SOLID WASTE COMPOSITION

Material	%
Paper	45.5
Organics	21.5
Metals	9.0
Stone, sand, other inerts	7.5
Glass	6.0
Textiles	4.5
Wood	3.0
Plaster, leather, rubber	3.0
	100.0

This classification is good but solid wastes should also be separated into
combustibles and noncombustibles. The noncombustibles can be classified
as glass, cans, metal, ferrous and nonferrous, and miscellaneous noncombustibles. Such analysis is needed in determining incineration feasibility.
Wet combustibles possess a heat value of 3500 BTU/lb, 6000 BTU/lb on a
dry basis. Domestic refuse will burn without auxiliary fuel, depending on
the concentrations of water and noncombustibles (Fig. 2). Heat values for
different wastes are often difficult to find. Essenhigh (1968) has given data

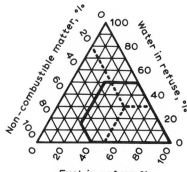

FIG. 2. Refuse will burn without booster fuel if its water and noncombus-
tible fuel contents are within limits shown.

on the typical composition of various solid wastes, percent moisture, combustible solids and the BTU values per lb (Table 3). Incineration is discussed separately.

Packaging materials are continually changing and increasing in volume.
Plastics, foils, fiber, and paper present the biggest problems. A glass bottle
in New York state costs 30¢ for pickup and disposal, seven times its original
cost. Incineration of plastics releases such tremendous heat values that older
furnaces are burned out. Plastics also require so much oxygen that loading to
incinerators must be reduced for efficient operation. Plastics also contain dangerous halogens, which are released to the atmosphere. In a city the size of
New York it costs more to dispose of the Sunday paper than it costs to buy it.

A. Sanitary Landfill

A sanitary landfill is not an open dump. It is a precise method of dispos-
ing of refuse to land without creating a nuisance or public health hazard by
utilizing principles of engineering to confine the refuse to the smallest practical area and to the smallest volume, and to cover it with a layer of earth at
the conclusion of each day's operation or more frequently if necessary.*

Four operations are used routinely in sanitary landfills: (a) solid wastes
are deposited in a controlled manner in a prepared portion of a specially selected site; (b) solid wastes are spread and compacted into thin layers; (c)
solid wastes are covered daily or more frequently with earth; and (d) the cover
material is compacted daily. Two general methods of landfill are used: the
area and trench methods. The ramp variation technique can be used with
either method.

The most important step in developing a landfill program is long range
planning and acquisition of suitable disposal sites. Geological factors are important. Other factors include: knowledge of public health and nuisance problems of uncontrolled disposal, standard operational procedures, capabilities
of equipment used, topography, climate, land use and future land use, access-
ibility, availability of earth cover, relation to residences and industry, average haul for collection vehicles, potential ground water contamination, zoning
regulations, and whether the site will gain public acceptance. Details on these
factors have been published by the American Public Works Association.**

*Committee on Sanitary Landfill Practice of the Sanitary Engineering Divi-
sion, Sanitary Landfill ASCE - Manual of Engineering Practice No. 39,
American Society of Civil Engineers.

**American Public Works Association, Municipal Refuse Disposal, Public
Administration Service, 2nd Edition, 1966, pp. 89-139.

TABLE 3

CLASSIFICATION OF WASTES TO BE INCINERATED

Classification of wastes / Type Description	Principal components	Approximate composition % by weight	Moisture content %	Incombustible solids %	BTU value/lb of refuse as fired	BTU of aux. fuel per lb of waste to be included in combustion calculations	Recommended Minimum BTU/hr burner input per lb waste
0 [a] Trash	Highly combustible waste, paper, wood, carboard cartons, including up to 10% treated papers, plastic or rubber scraps; commercial and industrial sources	Trash 100%	10%	5%	8500	0	0
1 [a] Rubbish	Combustible waste, paper, cartons, rags, wood scraps, combustible floor sweepings; domestic, commercial, and industrial sources	Rubbish 80% Garbage 20%	25%	10%	6500	0	0
2 [a] Refuse	Rubbish and garbage; residential sources	Rubbish 50% Garbage 50%	50%	7%	4300	0	1500

3	Garbage	Animal and vegetable wastes, restaurants, hotels, markets; institutional, commercial, and club sources	Garbage 65% Rubbish 35%	70%	5%	2500	1500	3000
4	Animal solids and organic wastes	Carcasses, organs, solid organic wastes; hospital, laboratory, abattoirs, animal pounds, and similar sources	100% Animal and Human Tissue	85%	5%	1000	3000	8000 (5000 Primary) (3000 Secondary)
5	Gaseous, liquid or semi-liquid wastes	Industrial process wastes	Variable	Dependent on predominant components	Variable according to wastes survey	Variable according to wastes survey	Variable according to wastes survey	Variable according to wastes survey
6	Semi-solid and solid wastes	Combustibles requiring hearth, retort, or grate burning equipment	Variable	Dependent on predominant components	Variable according to wastes survey	Variable according to wastes survey	Variable according to wastes survey	Variable according to wastes survey

a The figures on moisture content, ash, and BTU as fired have been determined by analysis of many samples. They are recommended for use in computing heat release, burning rate, velocity, and other details of incinerator designs. Any design based on these calculations can accommodate standard urban waste.

Geological characteristics of a landfill area must be determined from on-site testing and from soil surveys. Water tables, subsurface characteristics, and the direction and gradient of ground water flow must be known (Walker, 1969).

B. Garbage Grinding

One alternative in garbage disposal is grinding. It could decrease solid refuse by 10%. The savings in collection would only be slight. It could significantly reduce fly and rodent populations since both depend on garbage as a major source of food (Davis, et al., 1962). Average annual operating costs range from $17 to $22 per year. Sewage treatment costs go up where grinders are used by approximately 50 cents to $1.50 per capita per year.

C. Composting

The science of composting is actually an ancient one and was used for converting putrescible plant and animal residue to more stable materials for use as fertilizers and soil conditioners. Today it is used as a means of large-scale municipal refuse disposal (McGauhey and Gotass, 1955). Its greatest use was initially in Holland.

Composting is very similar in its action to a sanitary landfill. However, microbial reactions are controlled, yielding a more stable end product at a faster rate. Thus compost is one of the valuable resources available as a by-product of municipal waste disposal. It is also true that there is a real need to maintain the vitality and strength of our soils. Organic matter must be returned to the soil to do this. Composting removes the readily degradable organic matter from refuse and yields a stable material that can be used to reclaim waste land or to improve clay soils. Modern concepts of composting arose to a very large extent from the need to solve refuse waste disposal. They have been used by the small gardner and farmer for well over a hundred years.

The advantages of composting city refuse is apparent in many countries such as the Netherlands, where in 1961 it was reported that nearly 30% of all city refuse was converted to compost. This quantity is increasing every year. The process is also used in many parts of the world where the population per unit of land mass is very high, as in Japan, China, and India. Composting of refuse is practiced to recover nutrients and fertilizer, and also converting human and other animal excrement and plant trimmings to soil conditioning agents.

II. GENERAL NATURE OF CITY REFUSE
AND MATERIALS COMPOSTED

Refuse itself is perpetually changing in quantity and constitution. Decreasing density is a spectacular example of this. Increasing use of paper in packaging and increased use of plastics, tin, and waxes for liquid containers are all having an influence in changing the component factors of refuse. Thus, municipal waste varies in type, quantity, and constituent from city to city, from country to country, and from season to season. For example, city refuse is highly cellulosic, mostly because of the great amount of paper it contains. Also there is much more paper, rags, metal, and glass in American refuse than in European refuse. In tropical countries, during the monsoon season, the refuse becomes more saturated with water. Refuse in garbage consists primarily of organic material, water and insolubles.

Basically, the production of compost results from the breakdown of organic matter by microorganisms under aerobic conditions. Anaerobic composting processes are comparatively slow and odoriferous.

All types of solid organic matter can be composted under proper conditions. Waste sawdust from sawmills can be composted. Highly specialized microorganisms are involved and cellulose degradation in this process requires close control from both a pH and a nutrient standpoint. Cotton hulls from ginning operations have been successfully composted, as have corn cobs and other waste agricultural products. The major emphasis in the past decade has been on composting municipal refuse.

In 1952, analysis of a large amount of mixed municipal refuse was undertaken at Berkeley, California, by the University of California as a result of compost studies. The analysis gave the typical composition of refuse shown in Table 4.

TABLE 4

AVERAGE COMPOSITION OF MIXED REFUSE FROM BERKELEY, CALIF.[a]

	Amounts (%)
Physical composition	
Tin cans	9.8
Bottle and broken glass	11.7
Rags	1.6
Metals	0.9
Noncompostable waste of no value	7.6
Compostable material	68.4
Chemical Composition	
Moisture (as collected)	49.3
Ash (dry basis)	28.5
Carbon (dry basis)	35.7
Nitrogen (dry basis)	1.07
Phosphorus as P_2O_5 (dry basis)	1.16
Potassium as K_2O (dry basis)	0.83

[a]University of California Tech. Bull. 2, 1952.

There is a large amount of salvageable and compostable material in refuse, about 68.4% on a weight basis. One of the major problems in composting is the separation of the noncompostable material from the compostable material. The noncompostable material consists of 0.9% metal, 1.6% rags, 11.7% glass, and 9.8% tin cans. Recent equipment developments have greatly improved separation of these materials. The refuse can be ground to a coarse consistency and heavy metallic objects removed in a single operation. The larger pieces of glass which can be manually picked up are initially removed and the pieces that cannot be removed are pulverized to a fine powder and retained with the compostable organics. This can be removed in the final stage of composting by using screens and air to blow off the fines. Light ferrous metallic objects, such as tin cans, are separated by means of electromagnets. Other salvageable material such as rags are manually extracted. Today, the use of plastics in municipal refuse has increased to 2-3%, and these are also separated from compostable material before composting. Most plastics are nonbiodegradable.

Rags, scrap metal, and some of the cardboard and bottles are sold for salvage. When a market is available, as is the case on the Pacific coast and in the southwest of the United States, tin cans are sold and used in copper recovery.

The end product of composting is a material called "humus" whose chemical, physical, and biological characteristics vary because of differences in the nature of the raw material used and the conditions under which decomposition took place.

III. THEORY OF OPERATION

In the composting process the following factors are important:

1. Particle size of compostable materials.
2. Microorganisms in composting.
3. Temperature.
4. Air.
5. Moisture
6. Hydrogen-ion concentration.
7. Nutrients.

A. Particle Size of Compostable Materials

The size of the organic particles is very important in the rate of the composting reaction. In order to increase the surface area exposed for microbial attack, the particle size should be a minimum. Proper grinding is the key to efficient composting. This facilitates penetration and invasion of the organisms and permits oxygen to penetrate the organic wastes more readily. The natural materials such as lignaceous compounds, which physically block the penetration of organisms to the more readily available carbon sources, are also physically broken by grinding processes. Generally, rasps and pulverizers are used for this purpose, but with the Dano process this is partially achieved by attrition during the process itself. With proper mixing, acid and alkaline residues are neutralized and rendered more suitable for rapid biodegradation. This also ensures a more uniform product.

If the particles are too small, there will not be sufficient voids for maintaining an aerobic system. Grinding should be adequate to give a particle of

1/2 in. in mechanical composting. The particle sizes in windrow composting should be 1 - 1/2 in.

In some cases, two-stage grinding usually produces a more uniform material for composting than single-stage grinding--in the Dano composting plant at Sacramento, California, for example (Fig. 3). After separating the salvageable material from the refuse, the remaining refuse is mechanically conveyed to the primary grinder--a 25-ton Pennsylvania crusher. As shown in the flow sheet, the material that has been stabilized passes through a secondary grinder. This latter grinder makes a product easier to handle and with more uniform particles.

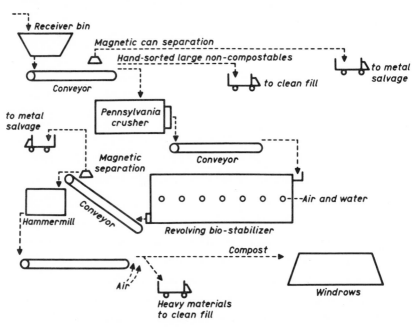

FIG. 3. Schematic flow sheet for Dano composting plant.

B. <u>Microorganisms in Composting</u>

The microorganisms responsible for the degradations of organics in the aerobic composting process are extremely numerous and diverse. Composting is carried out in a semimoist condition that favors the fungi and actinomycetes as the predominant groups of microorganisms, with bacteria assisting when moisture is adequate. Besides these, <u>Streptomyces</u>, and to a lesser extent algae, are also active in the composting. There is some evidence that the fungi, such as various species of <u>Penicillum</u>, <u>Rhizopus</u>, <u>Aspergillus</u>, and

Mucor, are the dominant organisms under dry conditions, whereas bacteria dominate under wetter conditions of composting. Both mesophilic and thermophilic organisms play important roles.

The mesophilic microbes function at temperatures up to 35-40°C, while the thermophilic group function at 50-65°C, with thermotolerant species functioning at intermediate temperatures. Very little work has been done on the microbiology of composting other than to show that natural mixtures of bacteria, actinomycetes, and fungi will produce a compost just as easily and just as fast as the so-called special pure cultures that some people feel are necessary.

In cases where, for one reason or another, the refuse is sterile, or in a near sterile condition, the use of processed compost for inoculum may hasten the composting process (Golueke and McGauhey, 1954). Since decomposition of the organics in composting is associated with the combined effect involving many species of microorganisms, inocula made up of a mixed population are more effective than one composed of a single species. Inocula containing mixed populations of microorganisms can be obtained from a number of sources including various animal and poultry manures, rich soils, other composts, and sewage sludge.

C. Temperature

Temperature is an important factor in composting. A thermometer is placed in the interior of the compost mass so the temperature can be determined at regular intervals. The optimum temperature in thermophillic composting is around 65°C, sufficient to destroy pathogens, vegetable seeds, and fly larvae. A compost reaching this temperature can be used without fear of disease organisms. The temperature cannot be allowed to go higher because of losses of nitrogen and inactiviation of beneficial microorganisms (Snell, 1960).

The primary microbial reaction is an aerobic thermophilic reaction. Sufficient air must be present to keep the system aerobic. Too much air passing through the composting mass will remove too much heat and the temperature will not rise to optimum. If there is too much moisture in the compost there will not be enough heat generated to raise the temperature to optimum. Thus, moisture is controlled closely with maintenance of optimum

temperature. When the temperature of the compost pile does not rise after
remixing, the compost can be considered as complete or at least past the
critical stage. In windrow composting the minimal time for killing pathogens,
e.g., Bacillus anthrax, Salmonella entirides, Erysipelothrix rhusiopathiae
(swine erysipelas), and Psittacosis-virus (parrot-fever virus), is 18-21 days
at a temperature of $55^\circ C$ or above. At slightly lower temperatures some
pathogens survive more than 251 days. Pathogens are killed more quickly
in shedded refuse.

Unground refuse processed in the Dano drum must be held five days to
kill pathogens. If a normal three-day drum retention time is used, the final
compost must be stored in windrows four days to kill the pathogens.

D. Air

Decomposition of organic material may proceed either aerobically or
anaerobically depending upon the availability of oxygen. Offensive odors,
which are difficult to control, are generated in anaerobic processes. Also,
anaerobic processes are extremely slow. Thus, modern composting must
be aerobic. A sufficiency of O_2 must be assured either by the turning of the
fermenting material during the process or by mechanical aeration. In mechanical composting, air is added to the composter continuously. It has been
found that $10-30 \, ft^3$ of air per day per pound of volatile solids are required
to maintain a good rate of biodegradation. There must be some excess air
supplied in order to ensure aerobiasis but not too much, as it will dissipate
the heat too rapidly.

The quantity of oxygen theoretically required by the aerobic composting
process can be computed from the stoichiometric relationship

$$C_a H_b O_c N_d + 0.5 \, (ny + 2s + r - c) O_2 \longrightarrow nC_w H_x O_y N_z + sCO_2 +$$
$$rH_2O + (d-nz) \, NH_3$$

where $r = 0.5 \left[b-nx - 3(d-nz) \right]$

$s = a - nw$

r and s are respectively equal to the number of moles of water and
CO_2 (Rich, 1963).

From the above equation, $C_a H_b O_c N_d$ and $C_w H_x O_y H_z$ are calculated on
an empirical mole basis. They express the compositions of organic material

initially and at the end of the process. Oxygen required by the process is

$$WO_2 = 0.5 \; (ny + 2s \mp r - c) \times 32 \times \text{(moles of organics entering the process)}$$

The rate at which oxygen is required varies with the type of organic material being composted and the temperature. Oxygen comsumption rates measured during aerobic composting of synthetic garbage at different temperatures within the range of $30 - 60°C$ were found to follow the relationship closely:

$$WO_2 = 0.07 \times 10^{0.31T}$$

where WO_2 is the rate of oxygen consumption, mg O_2/g initial volatile matter/hr

T is the temperature, $°C$

E. Moisture

Moisture content is very important to the biological activity in a composting mass. There must be sufficient moisture to soften the organic material to permit the microorganisms to hydrolyze the complex organic compounds into simpler compounds. Too much moisture results in lowered temperatures and anaerobic conditions, while too little moisture prevents metabolism. The optimum moisture content varies with the particle size and the physical characteristics of the organic material being composted. Generally, the finer the size of the particle, the lower the moisture content of the organic material should be. In the same way the coarser the particle, the higher the temperature should be. This is because particles too small in size resist gas exchange of O_2 and CO_2, and thus the voids tend to become filled with moisture, giving an anaerobic process. Generally, the range of moisture appears to be $30 - 70\%$; when the moisture content falls outside this range, composting proceeds very slowly. If the refuse being composted is too dry, water is added, and if it is too wet, dry material such as paper is added to absorb the excess moisture. In general, the optimum moisture content will fall within the range $50 - 60\%$.

F. Hydrogen-ion Concentration

The pH of the composting mass must also be carefully controlled. During the normal metabolism of microbes many organic acids are synthesized, lowering the pH unless sufficient buffering capacity is available. Lime is

generally added for pH control because of its low price, ease of handling, and
low solubility. Being moderately soluble, it is taken up by the acids as required. Since much of the metabolic activity is brought about by fungi, the
compost can operate at lower pH levels than other biological treatment systems. The opimum pH is 6.5, but satisfactory results can be obtained over
a pH range of 4.5 - 9.5.

G. Nutrients

The formation of microbial protoplasm, which is the net result of the
degradation of the organic matter in the refuse, requires a definite quantity
of nitrogen and phosphorous as well as trace elements. Nitrogen is a very
important factor in composting as it is in any media used to support biological activity. The nutrient requirements have generally been expressed in
terms of a carbon : nitrogen ratio. It has been found that an initial C:N ratio
(by weight) of 30:1 to 35:1 is optimum for aerobic composting. At greater
ratios, biological activity is impeded and composting time is increased. Below this range, nitrogen is in excess and ammonia is produced through deam-
ination of amino acids and by dentrification.

IV. PROCESS CLASSIFICATION

Two main types of composting are used: open windrow and mechanical.

A. Open Windrow System

This method of composting is performed in the open air. The refuse is
placed in outdoor piles approximately 4 - 6 ft high and 8 - 10 ft wide. The piles
are turned periodically to insure aeration and uniform composting. The moisture content of the compost is adjusted to approximately 60% and biological
activity is allowed to begin. After several days the waste heat from the microbial reactions will build up in the compost pile. The microbial reactions
will build up in the compost pile. The microbial populations shift from mesophilic to thermophilic reactions as the temperature often rises to $70°C$. The
system is maintained in an aerobic condition by turning it over before the
temperature rises too high. The moisture is also adjusted to 60%. Excessive
moisture from rainfall should be prevented by covering the piles and a water
percolator can be used to increase the moisture when it is too dry. Under
extreme conditions of dryness, sprinkling with water is required.

After turning the compost heap several times the temperature will fail to
increase back to the 60 – 70°C maximum commonly observed. The failure of
the compost to increase in temperature is used as a means of determining
when the compost has been stabilized. This open windrow composting process
requires approximately 6 – 10 weeks for completion. In dry weather it may
require only 2 – 3 weeks. It must be turned two or three times.

Composting conducted in piles in the open air can be operated satisfactorily if due attention is given to (a) maintenance of sufficient moisture, which
includes protection from saturated rain, and (b) turning and mixing to ensure
aeration. This method of composting is practiced in California and Israel
and on a large scale in the Netherlands. If the operation is small, the turning
over of refuse is completed manually. There are special mobile machines
for turning over the heaps at large composting depots.

In central America and at Kirkconnel, Dumfriesshire, the composting
depots use open-topped compartments containing a bed of mixed wastes about
4 ft deep on a perforated floor. This allows drainage of excess moisture and
admits air underneath to permeate the bed. After appropriate intervals, the
compost is transferred from one compartment to another by lifting the material
out with an overhead grab or a mobile shovel. At maturity the compost is
shredded and screened to exclude solids above 1/4 in. It is then sized and
bagged for sale. This method reduces the manual and mechanical operations
of turning over the heaps.

B. Mechanical System

When sewage sludge is collected already mixed with garbage and when
large quantities of garbage are to be treated, with or without sludge, it is
desired to conduct the first part of composting treatment in a special chamber equipped to control the required conditions. A special chamber with
mechanical devices has been constructed to turn the compost continuously,
adjust the moisture content, add air, inhibit growth of harmful organisms,
and speed up the composting process.

Some of the mechanical units have been constructed vertically, while
others have been built horizontally. Although the mechanical devices are
all proprietary, they operate on the same basic principles as the windrow
composting process. The only difference is the fact that the time required
to produce the stable compost is between 3 and 10 days.

A mechanical system can be operated to mix the contents of a special chamber either by (a) intermittent disturbance or (b) continuous disturbance. Intermittent disturbance is provided in chambers constructed vertically utilizing four to six special floors, one above the other. These floors can be opened and closed to transfer compost to the next lower floor. In some cases the floors open once every day, such as the Jersey process on the Island of Jersey in the British Channel Islands and the composting plant in Bangkok, Thailand. Both processes contain a six-floor fermentation chamber. The refuse, wetted by sludge, is elevated by conveyor to the top of the tower. The material is retained on one floor for one day. Thus at the end of six days the compost has reached the lowest floor level. The compost mass is held as a unit for two or three days and the heat increases to $60 - 70^{\circ}C$ to destroy all pathogens and seeds. The mass is then removed to the maturing sheds, where it is regularly transferred by an overhead grabbing device from cell to cell until the process is completed.

Another kind of chamber for intermittent disturbance utilizes a four-story grid-type construction over which an arm can rotate at intervals to cause the material to fall through the grid. The Eweson process in Zaragoza, Spain, utilizes this procedure.

The continuous disturbance system is carried out in chambers of cylindrical form that rotate, or in static systems equipped with ribbon screw-conveyors, which act to stir the compost. The former method is more commonly used. The Dano process, which uses a rotary drum, is employed extensively in Europe. There are over 52 plants operating in Europe and some parts of Asia.

The first plant was built in Denmark in 1933. The main features of this plant are shown in Fig. 3. It consists of a receiving hopper into which crude refuse is deposited. Refuse is then passed by elevating conveyor to vibratory or rotary screens with 1 1/4 in. perforation for the purpose of extracting dust and cinders into hoppers. These are situated to provide for gravity loading or large capacity vehicles for ultimate removal to the tip. After screening, the refuse passes over a picking table where salvageable items are manually extracted, and beneath magnetic separators, which discharge any separated tin into a hopper for baling. The refuse is then conveyed to primary grinders. The refuse leaves the grinder and is conveyed to an

elongated, slowly rotating drum known as the Dano "biostabilizer"--the main
component of the system (Scovel, 1958). The "biostabilizer" is a patented
apparatus and the patent is held by the Dana Corp. The drum is about 70 - 100 ft
long, 6 - 10 ft in diameter, and processes about 90 - 120 tons of ground refuse
per day. Although the size of the biostabilizer may appear formidable, it is
mechanically a simple piece of machinery with a minimum of moving parts.
Hence, breakdowns are rare and power requirements are low.

The biostabilizer provides an optimum environment in which microorganisms can break down the material. The drum is rotated at a speed of 0.8 rpm
during addition of refuse and at only 0.2 rpm during the remainder of the time.
The moisture content of the material is controlled by adding water through jets
located at intervals along the cylinder. Addition of water is not necessary for
liquid cannery wastes or refuse high in moisture. Controlled amounts of un-
heated air are provided by a manifold equipped with adjustable valves, so that
air can be introduced along the entire length of the revolving drum. As the
drum rotates, the refuse is carried forward until it cascades down. The refuse is retained in the drum for about 5 days. The composting material coming
out from the biostabilizer is conveyed to a vibratory screen, which removes
all material that does not pass through its 1/2 in. holes. The metallic material
that escapes the first removal is now removed by a second magnetic separator.
Again, the material passes through a secondary grinder, which gives a more uniform size of particles. Refuse is again separated through the use of a ballistic
type of separator dependent upon differences in relative densities.

This is accomplished by passing a jet of air through the compost as the
material drops from the end of the conveyor belt. Glass and other nonorganic
particles heavier than the compost are carried a shorter distance and hence
separated. Final maturation of the composting material is accomplished by
storing it in windrows in an open area for about 2 weeks. If storage area is
limited, turning the compost hastens this final stage. The last stage is characterized by a high incidence of actinomycetes and cellulose decomposing fungi.
The final volume of the compost varies from 60 - 70% that of the original fresh
material, and its weight decreases at a similar ratio. The Dano process is a
continuous one, refuse being fed in at one end and the composted material extracted at the other.

Another successful mechanical unit consists of a tier of rectangular composter cells, each of which is fitted with slowly moving slat bottoms. Refuse
introduced to the top cell is advanced by moving bottom from the receiving
end to the other end, where it is discharged by gravity to a receiving cell
located below. The refuse advances through the second cell and is then discharged in a like manner to a third cell. This process is generally carried
out in two tiers each consisting of three cells.

Another kind of chamber of continuous disturbance is a vertical cylinder
containing eight floors, with a rotating central shaft with rake arms at each
floor which move the bed of material, about 1 1/2 - 2 ft deep, to a hole at
the center or circumference alternately, so that the refuse flows to the floor
below and eventually out of the bottom of the system.

All the methods stated above are commonly used in composting. Before
an appropriate plant can be designed, all the relevant local circumstances
and the composition of the waste must be known.

C. Briquette Process

A composting process used in Germany is the briquette process (Brikol-
lane Vafahren). Dewatered sewage sludge is mixed with ground refuse in an
augor conveyor and conveyed to a briquetting machine. Briquettes 6 x 9 x 15
inches are made with a funnel on the underside. These are loaded automati-
cally in pallets and moved to a curing shed where the actual composting takes
place. Temperature in the briquettes increases to 60°C and fungi grow in the
surfaces. Moisture from the center of the briquette moves to the surface by
capillary action. This is adequate to sustain microbial growth as the water
level gradually decreases. Careful stacking permits air to move around the
briquettes and through the tunnels. Temperatures at the surface of the briquette reach 55°C. During curing, the moisture decreases from 63 to 13% by
weight. The briquettes are marketed to farmers. They may be easily broken
by a hammer mill.

Refuse high in paper processed by the Dano process possesses a composition of about 0.92% N, 0.57% P_2O_5, and a pH of 7.1. A typical compost
contains only 0.5% nitrogen, 0.4% phosphorous, and 0.2% potassium. Farmers
can buy and apply commercial fertilizers more cheaply than they can apply
compost or even livestock manure. Of course, the organic matter content is

important in improving the physical structure of the soil and in preventing erosion. This characteristic and its moisture bending properties are the prime reasons for composting (Fuller et al., 1960; Toth, 1960).

The C:N ratio of the completed compost will fall somewhere within the range 10:1 to 20:1. Compost that is not fully matured will have a tendency to remove nitrogen because of the excess carbon present, and for this reason it cannot be added to soil during periods of plant growth. The optimum C:N ratio is 20:1, while the C:P ratio is about 100:1. A nitrogen deficiency requires that an ammonium salt be added to the compost.

The finished compost has a dark or dark brown color characteristic of stable humus material. The pH value of municipal compost prepared by aerobic process finally ends up nearly neutral or slightly alkaline. The final moisture content of compost of biostabilizing or rotating drums processes is quite uniform, whereas that for windrows will depend upon the local climate condition, time of the year, and procedures employed. Normally, the terminal moisture content is about 25% or lower.

If, during the process, the operational precautions referred to are observed, foul odors will be prevented, flies discouraged, and pathogens destroyed. The resulting material will be of good quality and ready for use and even though it may not have great fertilizer value it is a most important soil conditioner. The rate at which compost is used depends upon the kind and condition of the soil and upon the crop that is to be grown. Forty tons/acre are used for poor soils.

V. ECONOMICS OF COMPOSTING

From the information available about costs of existing composting operations, some general guide lines are obtained regarding the economics of composting treatment.

The initial capital cost of constructing a completely mechanized composting plant is likely to be between $40,000 and $60,000 per ton of daily refuse input. Foundations and building will account for about 40% of the total cost, machinery and equipment for about 55%, and mobile plant for about 5%.

The labor force required will vary with the type of plant. Reasonable numbers are four or five men for a 20 ton/day plant, about 10 men for a 100 ton/day plant, and up to 18 or 20 men for a 300 ton/day plant, on the basis of one-shift operation.

Power consumption will also depend upon the type of plant, but is likely to range from about 8 kWh per ton of refuse treated for the intermittent disturbance method and about 15 kWh or more per ton for the continuous disturbance method.

Among the few composting processes that meet the demands of technical and economic feasibility is the Dano process. The cost of treating refuse by the Dano composting method is about $3.25 per ton, exclusive of salvage. Reported estimates of cost vary from $1.15 to $4.55 per ton.

Municipal compost seems to be a good source of available trace elements. This is particularly true in the case of copper, zinc, molybdenum, manganese, boron, and iron. Furthermore, the available trace elements are not present in excessive quantities and so the danger of specific metal poisoning of the crops cannot rise.

Certain wastes pose a problem in solid wastes disposal. These are hydrocarbons and industrial organics, demolition and construction rubble, wood (trees), leaves, automobile hulks, scrap metal, rubber and plastics, agricultural wastes, and sewage sludge. Methods of handling these wastes are quite varied. An acceptable procedure in one geographical area may not be acceptable elsewhere. Municipal and state authorities should be queried as to the methods being used in a particular locality.

VI. USE OF COMPOST

Mine land and land reclamation have the greatest potential use for compost. Anywhere soil erosion is high, compost requirements are high and use should be mandatory; 60% of the compost produced in Germany is used in this manner. Studies made on $30°$ slopes showed over 5000 gallons a year runoff. Addition of 79 and 159 tons of compost to adjacent plots reduced moisture runoff to 3000 gallons and less than 400 gallons respectively (Fig. 4). Soil loss for these same three plots were: no compost--150 ft^3/acre-year; 79 tons compost/acre--104 ft^3/acre-year; 159 tons compost/acre--less than 10 ft^3/acre-year (Hart, 1968). Loam and clay soils with 78 tons of compost/ acre will retain 2-15% higher moisture over the entire year (Fig. 5).

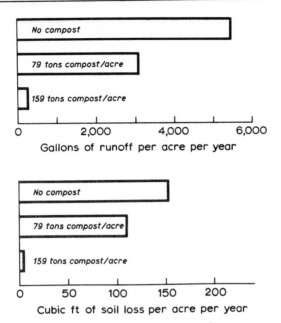

FIG. 4. Average water and soil loss from 30.1° vineyard slopes at Bad Kreuznach, Germany (after Hart, 1968).

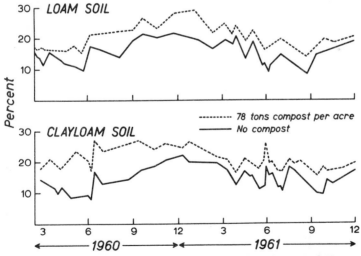

FIG. 5. Soil moisture regime on two vineyard soils at Bad Kreuznach, Germany (after Hart, 1968).

Composting has its greatest application in communities with populations of 10,000 - 1000,000. There are over 1760 such communities in the U. S.

Areas needing development are market research and market development, composting cost studies, and studies in waste land reclamation and pesticide degradation in high organic soils. In waste land reclamation the technology must be developed as well as the economics. Soil scientists, engineers, and economists are needed to solve the problems.

REFERENCES

Cohan, L. J. and J. S. Fernandes. 1968. "The heat value of refuse." Mech. Eng. 90, 47-51.

Davis, R. L. and R. J. Black. 1962. "Effects of garbage grinding on sewage systems and environmental sanitation." A. P. W. A. Reporter Dec. pp. 16-18.

Essenhigh, R. H. 1968. "Refuse rates in incinerators." Proc. of 1968, Natl. Incinerators Conference, ASME, American Society of Mechanical Engineers, p. 90.

Fuller, W. H. 1961. "Composting of city refuse." J. Milk and Food Technology 24, 385-389.

Fuller, W. H., G. Johnson and G. Sposito. 1960. "Influence of municipal refuse compost on plant growth." Compost Sci. 1, 16-19.

Golueke, C. G. 1960. "Composting refuse at Sacramento, California." Compost Sci. Autumn, 1960.

Golueke, C. G., and P. H. McGauhey. 1954. "A critical evaluation of inoculums in composting." Appl. Microbiol. 2, 1-12.

Gotaas, H. B. 1956. "Composting (Sanitary disposal and reclamation of organic wastes)." World Health Organization, Geneva, Switzerland.

Gotaas, H. B. 1960. "Material-handling methods for city composting." Compost Sci. 1, 5-9.

Grindrod, J. 1962. "Bangkok plant composts municipal refuse." Compost Sci. Winter, 1962.

Hart, S. A. 1968. "Solid waste management - composting-European activity and American potential." Report SW-2C-solid Waste Program U.S. Public Health Service, Contract 86-67-13. U.S. Govt Printing Office - 1968 - 0-309-787.

McGauhey, P. H. and H. B. Gotaas. 1955. "Stabilization of municipal wastes by composting." Trans. Am. Soc. Civ. Eng. 120, 897-901.

McKinney, R. E. 1962. Microbiology for Sanitary Engineers. McGraw-Hill, Toronto, Ontario.

Quon, J. E., A. Charnes and S. J. Wersan. 1965. "Simulation and Analyses of a refuse collection system." J. Sanitary Eng. Div., Prov. ASCE 91, SA5 Paper No. 4491, pp 17-35.

Rich, L. G. 1963. Unit processes of Sanitary Engineering. Wiley, New York.

Scovel, R. E. 1958. "The Dano Method of refuse disposal." California Vector Views, 5 6-9.

Snell, J. R. 1960. "The role of temperature in the composting process." Compost Sci. 1, 28-31.

Stirrup, F. L., and D. F. Gahan. 1961. "Constructing a new depot and disposal plant." Public Cleansing, 51, 6, 301.

Teenoma, B. 1961. "Composting city refuse in the Netherlands." Compost Sci. 1, 11-14.

Toth, S. J. 1960. "Using organic waste in agriculture." Compost Sci. 1, 10-14.

U.S. Department of Health Education and Welfare, Public Health Service. 1968. "Quad-City Solid Wastes Project, and Interim Report, June 1, 1966 to May 31, 1967." Quad-City Solid Wastes Committee Staff, National Center for Urban and Industrial Health, Cincinnati, Ohio.

University of California, Sanitary Engineering Research Laboratory. 1952. "An analysis of refuse collected and sanitary land fill disposal." Univ. Cal. Tech. Bull. 2.

University of California, Sanitary Engineering Research Laboratory, 1953. "Reclamation of municipal refuse by composting." Univ. Cal. Tech. Bull. 9.

Walker, W. H. 1969. "Illinois ground water pollution." J. Am. Water Works Assoc. Jan. pp. 31-40.

Wix, P. 1961. Town Waste Put to Use. Cleaver-Hume Press Ltd., London, England.

Chapter 12

SHOCK LOADING

Sewage collection and treatment works are generally designed and operated on the basis of handling municipal wastes that are of fairly uniform
quality and are received in quantities that follow a typical day-to-day pattern.

Since it is generally necessary to accept industrial wastes in municipal
works, it is important that the problem of shock load be recognized and
understood. Although the term shock load applies to all biological treatment
systems, it is generally recognized that the activated sludge process is
particularly susceptible (Gaudy and Engelbrecht, 1966), and most reports
relate to this area.

Shock loading is generally defined as any rapidly occurring or immediate
change in the chemical or physical environment. Shock loads are associated
with chemicals, temperatures, volumes, and/or masses of materials exceeding the limits of normal variation for which a treatment work is designed
(McTavish, 1963).

Shock loading may affect the established metabolic pattern in the aerator
or affect the flocculating and settling characteristics of the system. Essentially three major types of shock loads are encountered: (a) quantitative,
(b) toxic, and (c) qualitative.

The quantitative shock load, the most general concept of shock load, is
observed with an immediate name change in BOD loading. Figure 1 indicates
the normal variation and shock load areas of a standard activated sludge process. It should be noted that each variable has its own upper and lower limit
and that these limits are dependent upon time. A BOD concentration of

1,000 mg/liter may be tolerated for 5 min but not for 5 days in a plant designed for sewage having a concentration of 150 mg/liter.

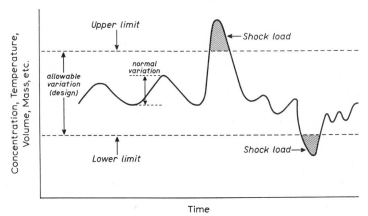

FIG. 1. Normal variation and shock loads in the activated sludge process.

High BOD shock load often causes sludge bulking. Bulking has also been attributed to the increased growth of Sphaerotilus and certain microbes (Eckhoff and Jenkins, 1966).

In practice it has been found that fluctuation in flow varies from about 50% to 150% of the 24-hr mean. Plant loadings are even more variable. Recent published figures indicate that the BOD loading can vary from as low as 10% to as high as 300% of the mean value even for large treatment facilities.

Figure 2 shows the variation in concentrations from the average for raw sewage and primary effluent from a medium-sized city in Ontario (McTavish, 1963). The raw sewage increased as much as 120 ppm above the average and decreased to the same extent below average. The primary effluent did not vary as greatly. The highest concentration was 60 ppm above average and the lowest 80 ppm below average.

To avoid this problem, organic wastes from industry should be regulated and wide BOD fluctuations balanced out. A maximum BOD concentration of between 300 and 500 mg/liter is suggested.

If a change results in a decrease in BOD which happens when excessive flow or widely fluctuating flow dilutes the sludge, the shock loading is termed hydraulic shock.

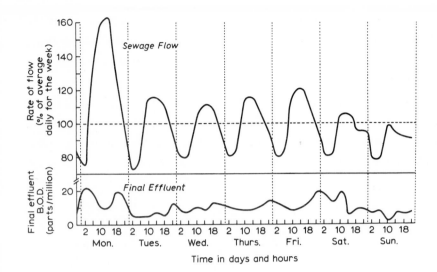

FIG. 2. Variation in the rate of flow and the final effluent BOD concentration.

In most sewage plants the flow reaches a peak during the day and becomes minimum after midnight. Flow rates during the week days will be greater than weekends and flows during the week days will be greater than weekends and flows during the spring months exceeds those of other seasons. High hydraulic loading results in "wash out" and the biological activity is lost from this system.

I. TOXIC SHOCK LOAD

Toxic shock involves an influx of organic or inorganic compounds which partially or totally damage the existing metabolism or disrupt the established environmental conditions of the microbial population.

Temperature is one of the most critical environmental factors affecting biological waste treatment system. Basic biochemistry has shown that the rate of enzyme reaction approximately doubles with each 10°C rise in temperature up to 35°C. As the temperature increases, the viscosity decreases and sedimentation is improved. A shock may be considered if there is a sudden variation of temperature from 30 to 50°C (ΔT). It is customary to place an upper limit of about 60°C on the temperature of industrial wastes discharging to sewers.

Rapid changes in pH of the waste are also considered to be in this class of shock loading. pH is easily controlled and is of less significance than shock loads caused by salts of heavy metals such as copper, zinc, chromium, and cadmium.

Since these materials are capable of affecting the microbial activity it is necessary to determine that concentration level of each ion which can be permitted in sewage plant effluent without upsetting the biological treatment. Suggested maximum concentrations for these materials are:

Cyanide as HCN	2 - 5 mg/liter
Copper as Cu	3 - 8 mg/liter
Chromium as Cr	3 - 10 mg/liter
Nickel as Ni	3 - 10 mg/liter
Lead as Pb	3 - 10 mg/liter
Zinc as Zn	3 - 10 mg/liter
Cadmium as Cd	3 - 8 mg/liter
Phenolic compounds	0.01 - 1 mg/liter
Sulfides as H_2S	2 - 5 mg/liter

Since most of the materials are quite toxic to biological systems high levels have to be avoided. Greases, oils, and radioactive waste are also considered as toxic shock loads.

The qualitative shock load is defined as a change in the chemical structure of the substrate and may constitute a serious type of system, particularly where joint treatment is practiced. It does not concern the total organic concentration, which indeed may remain the same as before the shock, nor does it imply that the change is toxic.

For example, change from carbohydrate waste to a proteinaceous waste, simple sugar to polymers, and sucrose to lactose may require a slight shift in the microbial population for optimal bio-activities to resume.

A. Biochemical Response to Toxic Shock Load

Based on the enzyme theory, methods have been developed by biochemists to describe the toxic effects of substances on purified enzymes. With this method it is also possible to differentiate between several types of enzyme inhibition, e.g., adaptive versus nonadaptive and competitive versus noncompetitive.

Very recent investigations in the basic fields have shown that several mechanisms of metabolic control may be operative in bacteria (Komolrit and Gaudy, 1966).

Malaney et al. (1959) describe several mechanisms by which the organisms recover:

1. The toxic nature of the ion is neutralized in some way by microbial activity, perhaps by oxidation to an insoluble oxide.

2. The initial ion-to-population ratio is high enough to inhibit oxygen uptake (respiration) but not multiplication of the organisms. Increasing numbers of bacteria reduce the ion to population ratio.

3. The bacterial population adapts to the use of the ion as a food material and thus removes the ion. Adaptive or mutant populations of microbes grow and replace the original population.

4. The oxygen uptake in the presence of the ion results from the growth of one species or a few species of microorganisms of the original population which were less affected by the ion than the majority of the population.

II. MINIMIZING THE EFFECT OF SHOCK LOAD

Adjustment of flow rate should be made. Combined sewer systems and those greatly affected by infiltration will periodically supply very high flows to treatment plants. Usually these flows occur over a short period of time.

If the plant is subjected to heavy flows and solids are lost in the effluent, the formation of a good activated sludge can be hastened by addition of ferric chloride (McTavish, 1963). The ferric chloride aids in coagulation and a good activated sludge require a small concentration of iron and aluminum slats to aid in good settling.

Many new plants are subjected to flows much below the plant design flow. This results in extended aeration times, nitrification, and very small floc particles, and the use of a large volume of air for each pound of BOD removed. Under these conditions only part of the plant should be placed in operation. Control of BOD of the feed is most important. Sudden and large changes in the BOD loading of a conventional activated sludge plant are often not dealt with effectively in plant design. Operations must be such as to minimize the following disadvantages to the conventional activated sludge process:

1. BOD loadings are limited to about 35 lb of BOD/day/1000 ft^3 of aeration tank volume.

2. High initial oxygen demand of mixed liquors is encountered.

3. There is a tendency to produce bulking sludge.

4. High sludge recirculation ratios are required for wastes with high BOD concentrations.

5. High solids loadings are imposed on final classifiers.

6. Air requirements per pound of BOD removed are high.

To deal with shock loading, operators must be fully acquainted with the design limits and capabilities of the treatment plant.

Many of the shock load problems such as flowrate, BOD, nitrogen deficiencies, etc., can be dealt with by slight modification of the conventional process. This may be necessary for short term operations until the source of the shock is removed or until the plant is enlarged. It may also be that the modification is the most effective and economical way of dealing with the problem on a long term basis.

Shock loads resulting from toxic ions, pH changes, and/or extreme temperatures usually require correction at the source.

REFERENCES

Eckhoff, D. W. and D. Jenkins. 1966. "Transient loading in the activated sludge process." Water Pollution Control Fed. J. 38, 3, 367.

Gaudy, A. F. Jr. and R. S. Engelbrecht. 1966. "Quantitative and
qualitative shock loading of activated sludge systems." Water Pollution
Control Fed. J. 33, 800.

Komolrit, K. and A. F. Gaudy, Jr. 1966. "Substrate interaction during
shock loading to biological treatment processes." Water Pollution Control
Fed. J. 38, No. 1 (Nov.), 1259.

Malaney, C. W., W. D. Sheets, and R. Quillin. 1959. "Toxic effects
of metallic ions on sewage microorganisms. Sewage Industrial Wastes. 31,
(11), pp. 1309, Quebec Water Board (personal correspondence).

McTavish, D. 1963. "Regional assistant supervisor minimizing the
effects of shock loads, n notes of the senior sewage works operators' course."
Ontario Water Resources Commission, pp. 79-88.

Chapter 13

SLIMES

A slime can be defined as an "accumulation of microorganisms and what-
ever other inorganic and organic debris may become imbedded in the mass"
(Shema, 1962). Slimes will vary in composition, depending on the diverse
physical, chemical, and biological factors of their environment (Crosby,
1953).

Sewage fungus, abwasser-pilz, heterotrophic biocoenosis, and slime infestation are all names for the community of organisms making up the slime (Curtis, 1969).

I. PHYSICAL AND CHEMICAL NATURE

It is important to determine the type of a particular slime deposit, because an analysis of its composition may serve as a basis for a decision on the type of control methods to be used.

In appearance, slimes are usually brown or brownish-grey in color for the first 3-5 days. In aerobic environments, the slimes can either remain brown or become green because of algal growth. If the dissolved oxygen content is high, slimes tend to be brown in color. If the content is low, they will be grey to black in color (Heukelekian and Crosby, 1956).

The type of slime growth and its consistency are related to the concentration of the medium in which the growth develops. Abundant and watery slimes grow in liquids with low BOD.

The odor of the slime is similar to that of the medium in which the slime is formed. The average value of moisture content is 94.7%, while the average value of ash is 24.1% of the dry solids.

A slime will grow on any surface, smooth or rough, provided there is adequate nutritive materials and a water-air interface. A slime may also grow as a free floating mass (Reid, 1953).

When attached slimes become detached they cause a secondary pollution problem. As the detached mass decomposes additional oxygen is removed from the water. Slime masses become detached because (Harrison and Heukelekian, 1958) (a) high river flows cause mechanical disruption of the attachment, (b) the microbial mass decomposes and becomes detached when the period of primary pollution ceases, and (c) the mass becomes too large.

In general, slimes are very similar in appearance and other characteristics to primary sludge. Typical slime manifestations occur in rivers or streams below areas discharging pollutants.

II. BIOLOGICAL CHARACTERISTICS

Slime is essentially a matrix of filamentous organisms (fungi, bacteria, and protozoa) which give the community a cohesive character and nonfilamentous motile organisms (protozoa, diatoms, and bacteria). Higher organisms (rotifers, nematodes, chironomids, larvae, etc.) are also present. These higher organisms probably feed off the slime rather than contributing to its formation (Curtis, 1969).

The relationship of the slime formation to the polluting source is shown in Fig. 1. As can be observed, slime is a relatively sensitive indicator of organic pollution in flowing waters.

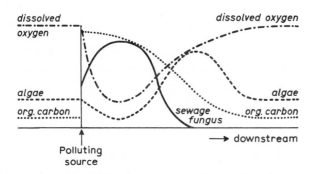

FIG. 1. Organic pollution in a river (after Curtis, 1969).

The most common organisms found in river slimes are (Curtis, 1969):

1. Bacteria -- Sphaerotilus natans, Zoogloea ramigera, Flavobacterium sp., Beggiatos alba.

2. Fungi -- Geotrichum candidum, Leptomitus lacteus.

3. Sessile ciliate -- Carchesium polypinum.

4. Algae - Stigeoclonium tenue.

5. Various diatoms.

The slimes found in pulp and paper mills, cooling towers, and sewers may or may not contain these same organisms.

Within any community of organisms there is at least one organism that exerts a major influence on the community because of its greater number, size, and activity. This is called the dominant species. The removal of the dominant species causes important changes, and any program to control the slime must be directed to the control of the dominant species.

Two ecologically dominant organisms are Sphaerotilus natans and Leptomitus lacteus.

A. Sphaerotilus natans

This bacterium is composed of individual cells united into unbranched filaments by a slime sheath surrounding the cells. In addition to the slime sheath and external to it, this bacterium produces a capsule or slime layer (Curtis, 1969). Reproduction is by fragmentation of filaments or by the formation of motile conidia. The relationship of this to other iron bacteria with detailed photo microgroups is presented by Zajic (1969). Different forms of the bacteria develop as a polluted water returns to normal.

Domestic and industrial sewage both contain the carbon compounds necessary for the growth of Sphaerotilus. The constituent compounds causing the BOD are more critical than the total BOD per se. It is for this reason that streams that are considered in excellent condition with respect to criteria such as DO and BOD support slime growth (Amberg et al., 1962).

Sphaerotilus is generally present in neutral waters, although Curtis (1969) has found that a pH range of 6.8-8.0 supports the best growth. Slime problems are more evident during the cooler seasons of the year. It is believed that at lower temperatures there is less competition for nutrients because of the lowered metabolic activity of other organisms. Also the biological treatment of effluents is not as efficient at lower temperatures.

River flow affects the oxygen supply and the supply of nutrients and also determines whether the slime remains attached. The oxygen supply is very important because Sphaerotilus natans is an obligate aerobe. It grows best in the 3-11 ppm oxygen range. Respiration can be determined using radiophosphorus (Varma and Reid, 1964). If the mineral composition is adequate, the amount of growth is proportional to the concentration of the available nutrients and the time interval over which the food is supplied.

Detachment of the slime masses is the cause of secondary pollution.

B. Leptomitus lacteus

Leptomitus lacteus is a member of the aquatic fungi belonging to the
order Saprolegniales. Filaments are 8-16 μ wide and approximately 400 μ
long. No cells are visible.

In a stream, the appearance of this fungus is similar to that of Sphaerotilus natans (Zajic, 1969). It differs from the latter by a coarse appearance
and the absence of mucilaginous flocs (Harrison and Heukelekian, 1958). It
grows in the pH range 2.5 — 7.5. It also is more evident during the colder
seasons and requires aerobic conditions.

III. OCCURRENCE

A. Rivers and Streams

A typical slime infestation occurs in a river or stream below a source
of domestic or industrial pollution. The occurrence may be sudden or
seasonal. It is esthetically undesirable and adversely affects the normal
river biota.

One of the difficulties with these slime infestations is that they occur
in streams that are considered in excellent condition with respect to DO and
BOD. For example, although the BOD in the Columbia River rarely exceeds
1 ppm, slime infestations are common (Amberg et al. 1962). Attached
slime growths have the ability to extract nutrients of low concentration from
large volumes of flowing water. Therefore, they present a very difficult
problem that cannot be solved with conventional waste treatment methods.

Once slimes develop in a river or stream they cause many problems.
In appearance the slimes are esthetically offensive. Slime growth on fish
eggs can prevent hatching (Curtis, 1969). Slimes are a nuisance to fisher-
men, fouling lines, clogging nets, even imparting an undesirable appearance
to fish. They foul ship bottoms, piers, and other structures in contact with
the infected water.

The secondary effects of slimes may be more serious than the primary
effects. When slime growths become detached, they are carried downstream
to a location where the stream velocity is low enough to allow them to settle
out. Here, the slimes start decomposing, and the odor plus the depletion of

the oxygen supply of the water causes problems. Secondary pollution is complicated by the fact that it occurs miles from the original source of pollution.

B. Pulp and Paper Mills

In pulp and paper mills there is growth and development of slime deposits in stock preparation systems, machine areas, and white water systems (Nelson, 1962). White water supports growth of microbes and
subsequent slime formation since it has a higher temperature and greater
nutrient value than fresh water (Spalding, 1960). The white water is continuously recycled, which induces the accumulation of slime.

In general, the problems resulting from slimes affect the operating
efficiency and the quality of the finished product.

The following are some of the problems caused by slimes in pulp and
paper mills (Spalding, 1960): (a) They reduce the effective diameter of
stock and water lines by sticking to equipment surfaces, (b) they contribute
to the plugging of wires and felts on the paper machine, reducing efficiency
and creating problems such as improper sheet formation, (c) their presence
in cellulosic pulp fiber affects the strength, appearance, and odor of the
paper produced.

Slimes also produce plugging of screens in the pulp and paper systems
and therefore reduce efficiency. When large amounts of slime become in-
corporated in the paper sheet, the strength is reduced. Sheet breakage
necessitates rethreading of the machine. Slimes may be responsible for
unsightly spots, holes, odors, and general discoloration throughout the
sheet (Pera, 1959).

The severity of a slime problem is determined by (Nelson, 1962):
(a) the type of buildup (i.e., its physical characteristics), (b) the rate of
buildup, (c) the location of the slime development. The rate of slime buildup depends upon (a) the type of pulp, (b) the efficiency of the pulp washing,
(c) temperature and pH changes, (d) aeration, (e) nature of the make-up
water, (f) the nature of the system (i.e., closed or open).

The problems created by the slimes are further complicated because
the slimy deposits and gelatinous coatings act to protect the organisms from
the adverse effects of heat, dryness, and chemicals. This makes the destruction of organisms embedded in slime extremely difficult (Spalding, 1960).

C. Cooling Towers

The water used in colling towers is recirculated, making it very prone to microbial growth as organics buildup in the evolving water. The resulting slime buildup causes fouling, reduces cooling by impeding the flow of water, interferes with the flow of heat, and causes other debris to collect (McKelvey and Brooke, 1959). As in most slime formations, the slimes in cooling towers are the result of excessive growth and development of a large number of species of algae, bacteria, and fungi (Shema, 1962). The use of sewage water as a resource for cooling water will cause larger buildup of slimes on tower surfaces.

D. Sewers

The following problems are caused by slimes in sewers (Crosby, 1953): (a) the production of H_2S with the resulting bad odor, (b) sewer clogging, (c) an increase of solids at the treatment plants, (d) the conversion of H_2S to H_2SO_4 by the sulfur oxidizing bacteria. The H_2SO_4 induces sewer corrosion.

IV. MEASUREMENT TECHNIQUES

A. Rivers and Streams

In stream pollution the objectives are (Wilson et al., 1960): (a) to locate the areas of slime growth, (b) to develop methods to correlate the areas of growth with various concentrations of nutrients and certain physical factors, (c) to determine the amount, condition, and distribution of detached, suspended slimes with special reference to adverse effects downstream from the growing areas. To help achieve these objectives, effective measurement techniques are necessary.

Two methods have been used to obtain quantitative data. The first method utilizes a net of 2.5-mm mesh mounted on a frame. This net is immersed across a stream and the floating slime mass is trapped, removed, and weighed. The second method is a modification of the above. It involves pumping the water from a definite depth. The slime growth is caught in an apparatus similar to a plankton sampler.

Wilson et al. (1960) developed or adapted existing models for use: (a) the bottom trawl, (b) the tile-box sampler, (c) The Reighard tow net.

1. Bottom Trawl

Slimes in streams tend to become weighted with silt, sawdust, and other materials, whereupon they readily sink to the bottom. To determine the distribution and characteristics of these slimes the bottom-trawl is used (Fig. 2).

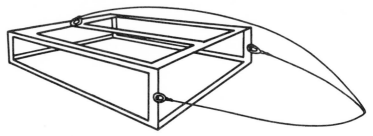

FIG. 2. Bottom trawl (after Wilson et al., 1960).

This device is 12 in. wide, 5 in. high, 20 in. long, and has sharp edges around the forward or open end. The top and bottom have metal panels, while the sides and aft have 20-mesh bronze screening. The weight is approximately 20 lb.

The device is thrown out as far as possible from a boat, and allowed to descend quickly to the bottom. With the boat in motion, the bottom is trawled for only about 3-6 ft. before the device is retrieved. Slimes settled on or growing from bottom surfaces are dislodged and collected in the trawl.

2. Tile-Box Sampler

This apparatus (Fig. 3) is used to provide a suitable attachment surface for the slimes. It also acts as a means for quantitative determination of the rate and volume of growth.

The sampler accommodates four 4 1/4-in. tiles. It is oriented in the current so that the water passes freely through the open box, past the tiles which are held in a vertical position edgewise to the flow. Provision can also be made to suspend microscope slides. In order to suspend the sampler in a level position a ballast may be required.

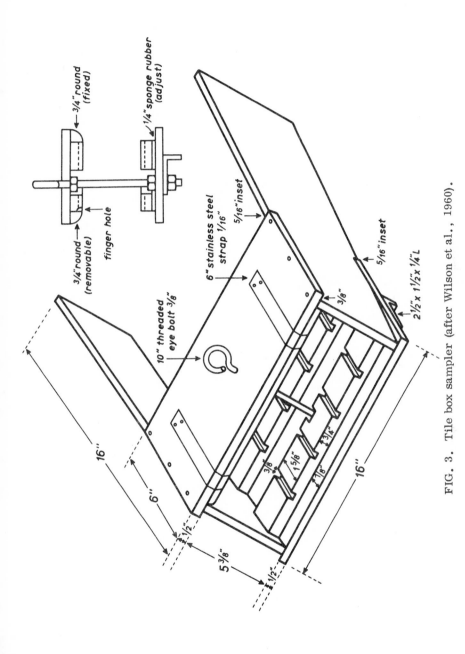

3/4"round (fixed)

1/4"sponge rubber (adjust)

3/4"round (removable)

finger hole

6" stainless steel strap 1/16"

5/16"inset

5/16"inset

2½" x 1½" x ¼" L

3/8"

10" threaded eye bolt 3/8"

16"

6"

1½"

5 3/8"

1½"

16"

3/8

1 5/8"

1 1/8"

1 3/4"

FIG. 3.   Tile box sampler (after Wilson et al., 1960).

After the box has been held in the stream for the desired period, often amounting to one to two weeks, it is slowly lifted to the surface. The samples collected on the tiles are ready for observation, laboratory analysis, and photographing.

3.   Reighard Tow Net

The Reighard tow net is used to determine the quantity, quality, and distribution of detached slimes (Fig. 4).

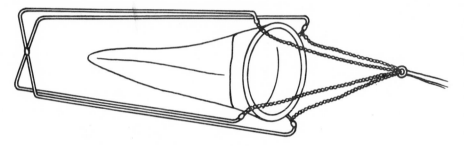

FIG. 4.   Reighard tow net (after Wilson et al., 1960).

The area of the net opening is 1 ft^2 and the silk bolting cloth used consists of 16-mesh netting. The duration of the sampling is timed, and the stream velocity is determined. Since stream velocity, duration of sampling, net area, and the quantity collected are all known, it is possible to estimate the amount of detached slime in the stream.

B.   Pulp and Paper Mills

In a pulp and paper mill the measurement of slime growth accretion is also important. The simplest and oldest device used for this purpose is a series of wooden strips suspended at conveniently accessible points in the mill system. The thickness of slime accretion on these strips is measured. This device gives a good indication of the progress of slime control.

Appling (1955) describes two units that are used for slime measurements. The first consists of a pine board protected by a perforated stainless-steel cage. This unit is submerged in white water at a specific location in the system. Every 24 hr the slime accretion is scraped off and the wet weight obtained.

The second unit is a box with a sample panel in it. This box is connected to any portion of the mill system. Every 24 hr the slime is scraped off the panel and the weight weight obtained.

In general, measurement techniques may give the overall picture of occurrence of the growth of slimes, but they do have limitations as follows:

1. Fragile growths such as fungi are easily detached, and therefore it is not possible to get an accurate measure.
2. The growth collected can contain much debris.
3. It is impossible to distinguish between living and dead material.
4. In order to get results that are related to the amount of pollution, it is necessary to take continuous samples and to have knowledge of the organic loading.

Laboratory analysis for slime formation is far more difficult. There are really two basic reactions involving slimes. One relates to the nutrient absorption behavior of films on biological filter media and the "schmutz-decke" or slime layer penetration into sand filter media. In the first instance, food is absorbed by the film, after which it is assimilated, resulting in a reduction of BOD in the fluid passing through the filter. In the second reaction, after the slime has coated the media, adhesion of suspended solids to the slime coatings takes place. This latter phenomenon is so important in water filtration that slime coatings are often promoted by deliberate insertion of floc into the "in-depth" filter input lines, while in the former case strenuous efforts are often made to regulate film thickness for optimum BOD reduction.

Even though slimes may have a use in filtration, excessive buildup in the event of delayed backwashing phases can lead to serious plugging problems, especially in the popular rapid sand filter. Slimes are also a serious problem in water softeners and can be dangerous to health unless suppressed by chlorine or other disinfectants. This latter problem is often a severe hazard to purity in industrial water conditioning systems.

Crosby (1953) ran a series of tests to determine the nature of slime behavior in sewage and refinery wastes and produced some interesting results from the point of view of designing equipment to encourage film growth. The more important of his findings for the present purpose are summarized here. (a) Negligible changes in film formation were found due to surface-

to-volume ratio or the presence of biological inhibitors in the coatings of
test material. (b) Speed variation over the range 0.40-1.48 ft/sec showed
a retardation of growth ratio of about 4:1 in the highest part of the range.
(c) Aeration was the most significant factor in stimulating growth in all environments, while immersion depth of test plates was mainly significant for
the indirect effect of oxygen level, i.e. the decrease in oxygen present at
increased depths was responsible for decreased film growth.

The effect of relative speed variation between seawater and ship's hulls
in marine fouling was noted by Zobell (1939). Normally, fewer fouling organisms will attach to panels in turbulent water. Vessel cruising speeds influence attachment growth, and vessels foul more readily at anchor in bays
where bacterial nutrient level is relatively high and where the speed between
the water and the vessel is at or near zero.

Reid and Assenzo (1960) described a slime testing unit comprised of
several glass drums of 2-in. width and 5-in. diameter to study slime formation. Tests showed that optimum film formation occurred at a pH of 7.2 and
a temperature of 27°C (81°F). The roller speed was 2 rpm.

A bioslime evaluator is shown in Fig. 5. It can be seen that liquid feed
comes from a reservoir on top of the main tank, moving by gravity through
a small flow meter and down over the heater into the bath in which the rolls
rotate. Temperature control is achieved by a small thermostat and heater.
Chemicals for the adjustment of pH, nutrients, replenishing liquor, etc.,
can all be added directly to the reservoir. Main bath temperature is taken
from a thermometer attached to the side of the tank while pH readings and
other routine samples can be taken either from the reservoir or the sump
at the rear.

A unit of this type permits evaluation of (a) the effect of gaseous atmos-
pheres on the control of slime, (b) the effect of various bedding materials
and roughness, (c) filter behavior, and (d) waste composition. A gas-flow
meter is located on the inlet side near the fluid-flow meter, while the discharge outlet is located at the opposite end. The unit is not constructed to
withstand pressure although this feature could be added. Rollers should be
constructed out of the desired test material. For routine purposes at least
four types of rollers are required, e.g., copper, Plexiglass, stainless steel,
and wood. The rollers are constructed so they can easily be removed and

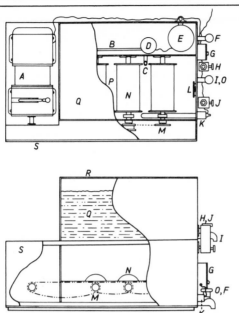

FIG. 5. Bioslime evaluator. A, 30:1 Zeromax speed reducer; B, float arm and pump microswitch; C, main bath overflow and liquid seal; D, encapsulated microswitch; E, circulating pump (0.1-1 gal/min); F, sump pump drain valve; G, main switch box; H, gas flow meters (1-5 SCFH); I, reservoir drain valve; J, fluid flow rotameter (10-40 ft^3/hr); K, main bath liquid heater and thermostat (150 W); L, wall mounted thermometer (4-80°C); M, sprocket and chain drive; N, test roll assembly (eg. steel, plastic, copper etc); O, bath drain valve; P, main bath; Q, reservoir; R, removable reservoir cover; S, gas tight main cover; Not shown: float control, strainer, pump sump rear wall, and float control guides.

replaced. Reid and Assenzo (1960) have found that a rotating drum device produces results that compare with actual values obtained in the field from biological filters. For optimal sorptive capacity, media used in trickling filters normally require a film thickness of 0.25 mm. Film thickness in this unit can be regulated by rotation and by adding a doctor blade.

V. METHODS OF CONTROL

In choosing a slime control agent, the following factors should be considered (Pera, 1959): (a) cost of the agent, (b) quantity of the agent required, (c) its effectiveness in any particular application, (d) its persistence or ease of removal from the finished product, and (e) its continued effectiveness over long periods of use.

The primary purpose of a slime control agent is to kill or inhibit the growth and accretion of microorganisms. In order to obtain adequate control of microorganisms in water systems it is necessary to use slime control agents that are effective against a wide variety of microorganisms. It is important to realize that high concentrations of agent will kill, lower concentrations will inhibit, and that extremely low concentrations may actually stimulate the growth (Shema, 1962). Effectiveness of an agent is a function of its concentration and time.

Chlorine is the most universally used biocide. It is highly toxic to algae, fungi, and bacteria. However, since it is an oxidizing agent, it must be added in sufficient quantity to satisfy the reducing of other impurities in the water before it is completely effective. For this reason a residual chlorine level is required. The residual will depend upon the system, the resistance of the organisms, and the location of the problem.

One part per million of chlorine for 30-min exposure or 0.5 ppm for 120-min exposure prevents the growth of slime, while 50-100 ppm are required to remove established slimes (Curtis, 1969). If the pH of the system is below 6, the addition of chlorine may cause corrosion of metallic equipment (Appling, 1955).

A. Rivers and Streams

Ideally, control of slimes in rivers and streams is achieved by preventing the entry of any substances that promote the growth of microorganisms forming the slime. Since this is impossible in most instances, other control methods have had to be improvised.

Amberg et al. (1962) report a control method in practice on the Columbia River. As a result of a survey of the Lower Columbia River in 1941 and 1942, it was concluded that Sphaerotilus natans grew only in areas where the concentration of spent sulfite liquor was 50 ppm or greater. On the basis of this survey, the Crown Zellerbach mill on the Columbia River spent over $200,000 in 1950 to install a deep water diffuser line which discharged the effluents to a depth of about 45 ft. This diffuser was effective in reducing the waste concentrations to less than 50 ppm, but the slime growths were not sufficiently reduced to eliminate the problem.

In laboratory studies by the National Council for Stream Improvement, it was found that (a) a 2-hr discharge followed by a waste storage period of 22 hr was effective in eliminating the slime growth, and (b) a discharge interval of 12 hr/day, followed by a waste storage period of 12 hr was not effective in controlling slime growth.

Two objections might be raised against the method outlined in (a) above. A discharge interval of 2 hr followed by 22 hr of storage would result in a 12-fold increase in waste concentration in the effluent. Also, under some circumstances the short interval (22 hr) might cause merging of the discharge downstream.

In an attempt to overcome these objections it was decided to increase the storage period and the discharge interval. The results of a laboratory study of this plan are shown in Fig. 6. It is interesting to note that slime growths could not be established by intermittent feeding unless there was a pregrowth already present.

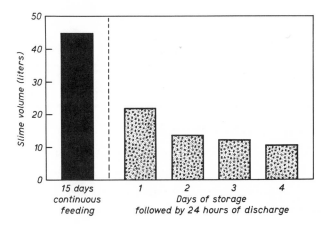

FIG. 6. The effect of intermittent discharge (after Amberg et al., 1962).

An experiment was conducted in the Columbia River using groups of logs as attachment surfaces for the slime. These surfaces were individually exposed to either continuous or intermittent feeding of spent sulfite solids. This experiment was conducted for about 5 months. The surfaces exposed to the continuous feeding had considerable growths of slime on them, and the mass of these growths was directly proportional to the concentration of the effluent. The surfaces exposed to intermittent feeding had the least growth of slime. On the basis of these experiments, a full-scale program was developed at a cost of $1.25 million for slime control (Fig. 7) at the Crown Zellerbach mill.

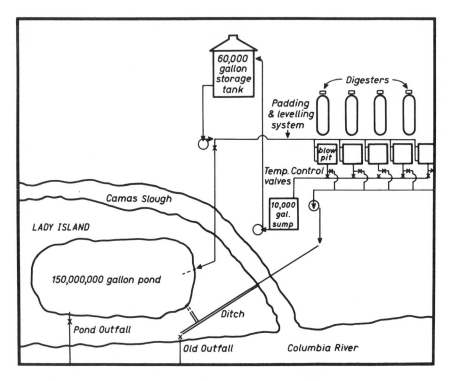

FIG. 7. Disposal system (after Amberg et al., 1962).

The system is capable of collecting either just the "strong wastes" or the total effluent. These can then be stored for intermittent discharge. The system has the capacity to store total effluent for 6 days or "strong wastes" for 60 days.

In 1961, impoundment of the total effluent began. It was decided to discharge for 24 hr/week. About 17 x 10^6 gal/day of effluent were discharged to the holding pond, and on the seventh day about 112 million gallons of dilute effluent were discharged to the stream.

Observations during the spring and fall of 1961 showed that the river was relatively free of slime (a reduction of 87%). It appears that continuous, extremely low concentrations of BOD will support slime growth, whereas intermittent high concentrations of BOD will not support its growth.

It should be pointed out that even during periods of low flow and high temperatures, the intermittent discharge did not lower the DO level below 7 ppm. No report was made on the BOD reduction in the waste resulting from the six-day retention period, which could also improve the operation.

B. Pulp and Paper Mills

The methods of slime control now employed in pulp and paper mills include (Spalding, 1960) (a) mechanical and chemical treatment of mill equipment when the mill is not in operation, and (b) the addition of biocides to the circulating water system during the mill operation.

Some of the biocides that have been used in the pulp and paper industry include chlorine, chloramine T, copper sulfate, organomercury derivatives, polychlorophenols, quaternary ammonium compounds, and combinations of the same (Table 1). Lederer and Delaney (1960) state that phenylmercurials are the most widely used control agents. Some organisms have the capability to build up an immunity to the various compounds used (Spalding, 1960); therefore, treatment methods must frequently be altered.

1. Organo-mercurials

Lederer and Delaney (1960) investigated organomercurials as slime control agents and found di(phenylmercuric)-ammonium propionate to be very effective. Di (phenylmercuric) ammonium salts result in a product 2000 times more soluble in water than that obtained with phenylmercuric salts. This higher solubility allows the entire strength of the slimicide to be made available, giving more efficient and economic control than continuous small additions of phenylmercuric salts.

These di(phenylmercuric) ammonium salts were tested on Aerobacter aerogenes, Bacillus mycoides, Aspergillus niger, and Penicillium expansum, which are believed to represent the types of organisms most commonly associated with slime problems in pulp and paper mills. Of all the di(phenylmercuric) ammonium salts tested, the acetate, propionate, and methacrylate are the most effective slime control agents. Further tests showed that

TABLE 1

BIOSLIME INHIBITING COMPOUNDS

Compound or Chemical	Experimental results to date	Range of Dosage, ppm
Calcium hypochlorite ("Perchloron")	Good	0.1 - 2.0
Sodium hypochlorite	Good	0.1 - 2.0
Copper sulfate	Fair	1.0 - 10.0
Copper napthenate	Fair	12.5 - 100
Sodium pentachlorophenol	Poor	0.7+
Chlorine	Best	0.05 - 5.0
Copper oxide	Poor	
"No-Slyme" (powder)	No data	–
Rosamine acetate	No data	–

Note: A solution of 5% caustic with 1% sodium sulfite and a wetting
agent applied at $110°F$ has also been found effective for cleaning
out slime coated vessels such as cation exchangers.

the propionate was the best. It has a high antimicrobial activity, high water
solubility, and a marked resistance to deactivation.

Mill trials with salts of propionate showed that the di(phenylmercuric)-
ammonium propionate is a very effective slime control agent. In addition
the compound is effective as a slime control agent even after 8 weeks of continued treatment, i.e., these organisms had not become resistant to it. In
comparison with a phenyl-mercurial-chlorinated phenol mixture (Product A)
and a nonmercurial (Product B), the di(phenylmercuric) ammonium propionate is more effective and more economical as a slime control agent than
product A or B (Figs. 8 and 9).

Thus it appears that di(phenylmercuric)-ammonium compounds have
greater potential as slime control agents. Appling (1955) states that care
must be taken in the use of organomercurials since high concentrations in
contact with brass or copper equipment will cause corrosion.

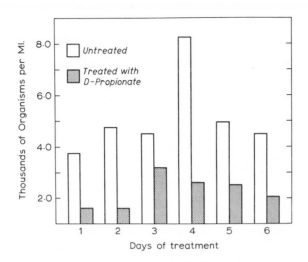

FIG. 8. Evaluation of D -propionate (after Lederer and Delaney, 1960).

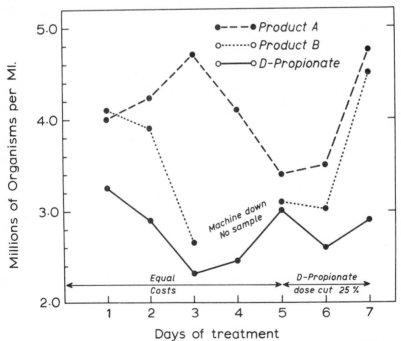

FIG. 9. Organisms 1 hr after addition of slimicide (after Lederer and
Delaney, 1960).

2. Bis-tri-n-butyl tin oxide

Connolly (1957) describes tests performed on the slime control agent bis-tri-n-butyl tin oxide, which is available from the Metal and Thermit Corporation under the registered trademark TBTO.

This compound is effective in the control of slimes caused by micro-organisms in paper mill water systems. It is less toxic than most other slimicides and perhaps for this reason does not show any corrosive effects on the wire or metal parts in a mill. The oxide should be added as a concentrated material. It has been found to perform efficiently under all pH conditions.

This compound has been tested in Canadian and New England paper mills and found to be very effective in the control of slime. In Canada it was compared with phenylmercuric acetate, while in New England it was compared with chlorinated phenols. In both cases TBTO was found to be more effective and cheaper.

TBTO has good adhering tendencies and therefore is able to control slime growth on the walls of tanks, pipes, etc. It also serves as a preservative and protects both the pulp and the finished product against mildew.

3. Methyl 2, 3-dibromopropionate

Cruikshank (1964) shows the results of using a methyl 2, 3-dibromopropionate. When the use of an organosulfur product failed to control the microbial activity in a pulp and paper mill, methyl 2, 3-dibromopropionate was tested. During this experiment the dosage of a 30% active formulation varied from 0.3 to 1.5 lb/ton of paper pulp.

It was found that this compound kept the white water system essentially free of slime and the microbial population under control. This test was continued for three weeks and excellent control of both bacterial and fungal types of microbes was obtained.

4. 2 Allyl-chlorophenol

Spalding (1960) describes the use of compounds from the group consisting of 2-allyl-3, 4 dichlorophenol, 2-allyl-3, 4, 6-trichlorophenol, and their salts for slime control. The use of 50-500 ppm (by weight) of these compounds

effectively reduces the slime-forming microorganisms in the aqueous paper--
making medium. These compounds are effective in the pH range 4.5-9.5.

5.   Cyanodithioimidocarbonate with Ethylenediamene

The use of a solution in which sodium cyanodithioimidocarbonate is combined with an equal proportion of ethylenediame for slime control is proposed
by Pera (1959). This solution was tried at a paper mill where there were
many slime problems. Previously, it had been necessary to shut down the
machines as often as twice per 24 hr to remove slime. The addition of 1-3
ppm of this solution resulted in a slime-free operation for periods of three
days and longer.

No chemical compound is equally effective against all types of microorganisms or under all types of operating conditions. To assist in the selection of the proper biocide, an evaluation of toxicants has to be performed.
The purpose of an evaluation is to separate the good toxicants from the poor
ones and to test the most promising ones in mill tests (Appling, 1955).

Finally, it must be stated that good housekeeping is probably one of the
most important slime control methods and should be practiced in every mill.
Proper cleaning of the equipment with a toxicant-detergent can be very effective in the control of slimes.

C. Cooling Towers

Fouling in cooling towers (McKelvey and Brooke, 1959) is caused by:
silt of other suspended solids in the makeup water, airborne dirt or other
particles carried into the tower by the incoming air, substances such as oil
leaking into the recirculating water, the deterioration of the cooling tower
timber, biological growth within the tower. Fouling caused by biological
growth is the most serious type and the most difficult to combat.

Before any type of control is recommended for slime prevention, an
analysis of the cooling water, the raw water supply, and the organisms in the
slime should be completed.

Some of the common methods of slime control in cooling towers are reviewed by Talbot et al. (1956). The use of chlorine is a fairly common control method. A residual of 0.5 ppm will destroy most microorganisms, and
no immunity is built up. Continuous feeding is better than intermittent feeding

(McKelvey and Brooke, 1959). Bromine gives good results in general, but because it is expensive and dangerous it is neither economical nor practical. The combination of continuous chlorination with shock bromine treatments gives good results at a cost lower than either of the two used separately. Good results have been obtained with copper sulfate. The average dosage ranges from 0.1 to 0.5 ppm. One problem with it is that it precipitates at a pH of 8.5 or over. Also, if the treated water subsequently enters a stream, it is necessary to ensure that the concentration of copper ion is below 10 ppm because of its toxicity to aquatic life.

Sodium pentachlorophenate is soluble and stable in alkaline water. It is usually slug-fed with enough chemical added to establish a concentration of about 200 ppm. Satisfactory results have been obtained by combining the phenolic compound with copper salts. Care must be taken when the treated water is discharged into a stream that the aquatic life is not killed. It is also important to ensure that this biocide is compatible with the corrosion inhibitor and not corrosive to metals (Shema, 1962).

Quaternary ammonium compounds are seldom used alone because of the high concentrations required. They do possess effective surface active properties and have been used with other algacides to assist in the attack on the growth of algae.

Experience in a generating station in Providence, R. I., has shown that hypochlorination is an effective method of slime control in cooling towers (Springs, 1957). Previously, gaseous chlorine was used to control slime in the surface condenser cooling water. The residual was at 0.5 ppm and the feed was intermittent. With this system it was found that the chlorinators required a lot of attention and the hazards associated with using chlorine gas were always present. The chlorinators were replaced by hypochlorinators in 1955. The daily cost of treatment was reduced.

It was found that the sodium hypochlorite is as effective as chlorine in the control of slime. If storage facilities are included, the cost of installation of a hypochlorinator is lower than that of a chlorinator.

D. Sewers

The problems created by sewers are generally considered of minor importance conpared with those in the pulp and paper industry or in cooling towers.

Some control of slimes in sewers could be achieved by (a) constructing sewers at a grade that would ensure relatively high velocities and good scouring (the high velocities would prevent serious slime growths), and (b) intermittent cleaning of sewers either by mechanical means or by the use of fire hoses.

VI. USE OF BIOLOGICAL SLIMES

Up to this point the problems of slimes and their control procedures have been discussed. Biological slimes also have a variety of uses.

A. Trickling Filters

In a trickling filter the removal of organic material from waste water is accomplished as a result of an adsorption process that occurs at the surfaces of biological slimes covering the filter media. Subsequent to their removal the organics are utilized by the slimes for growth and energy.

As the liquid waste passes through the filter it contacts the slimes in thin films. Concurrent with the sorption of organics from the liquid onto the slimes, inorganics resulting from the oxidation of previously adsorbed organics are discharged to the liquid. Also, oxygen diffuses from the atmosphere into the slime, and carbon dioxide is discharged to the air (Rich, 1963).

B. Source of Polysaccharides

Although slimy deposits of Sphaerotilus natans are common in streams, industrial plants, and sewage treatment plants, the chemical composition of this slime has been studied by only a few researchers.

Gaudy and Wolfe (1962) in their studies performed chromatographic and chemical analyses of the capsular polysaccharide of Sphaerotilus natans. They found the major components to be fucose, galactose, glucose, and glucuronic acid.

Considering the masses of slime that can form in various locations, it is possible that Sphaerotilus natans could be used as a source of polysaccharides and used as a polyelectrolyte type flocculant.

REFERENCES

Amberg, H. R., J. F. Cormack, and M. R. Rivers. 1962. "Slime growth control by intermittent discharge of spent sulfite liquor." TAPPI, 45, 770-779.

Appling, J. W. 1955. "Slime in mill systems and their control. Microbiology of pulp and paper." TAPPI Monograph, No. 15, 97-133.

Connolly, W. J. 1957. "A new slime control agent." Paper Trade J. 141, 46-47.

Crosby, E. S. 1953. "Slime growth on submerged surfaces in sewage and refinery wastes." Thesis, Rutgers University, pp. 1-259.

Cruickshank, G. A. 1964. "Slime-controlled industrial process water system and process." U.S. Patent, 3, 151,020, (Cl. 162-190).

Curtis, E. J. C. 1969. Sewage fungus: Its nature and effects." J. Int. Assn. Water Pollution Res. 3, 289-311.

Gaudy, E. and R. S. Wolfe. 1962. "Composition of an extracellular polysaccharide produced by Sphaerotilus natans." Appl. Microbiol. 10, 200-205.

Harrison, M. E. and H. Heukelekian. 1958. "Slime infestation - literature review." Water Pollution Control Fed. J., 30, 1278-1302.

Heukelekian, H. and E. S. Crosby. 1956. "Slime formation in polluted waters II factors affecting slime growth." Water Pollution Control Fed., 28, 78-92.

Heukelekian, H., and E. S. Crosby. 1956. "Slime formation in sewage III nature and composition of slimes." J. Water Pollution Control Fed., 28, 206-210.

Lederer, S. J. and W. J. Delaney. 1960. "Di (phenylmercuric) ammonium propionate - a new slime control agent." TAPPI, 43, 160-166.

McKelvey, K. K. and M. Brooke. 1959. The Industrial Cooling Tower. Elsevier, New York. pp. 227-254.

Nelson, J. R. 1962. "Recent emphasis on dispersion concept in slime control." Paper Trade J. 146, 42-45.

Pera, J. D. 1959. "Processes for the control of slime-forming and other microorganisms and compositions for use therefore." U.S. Patent, 2,881,070, (Cl. 92-3), 10 pp.

Reid, G. W. 1953. "Sewage slimes." Purdue University Engineering Bulletin, 96, 476-482.

Reid, G. W. and J. R. Assenzo. 1960. "Biological slimes" in Biological Waste Treatment." Ed. Eckenfelder, W. W. and J. McCabe. Reinhold, New York, p. 347.

Rich, L. G. 1963. Unit Processes of Sanitary Engineering. Wiley, New York, pp. 47-49.

Rydholm, S. A. 1965. Pulping Processes. Interscience, New York, pp. 277-362.

Shema, B. F. 1962. "Microbiological control." Proc. Am. Power Conf., 24, 788-795.

Spalding, D. H. 1960. "Method of controlling slime by treating with a 2-allyl-chlorophenol." U.S. Patent, 2,922,736, (Cl. 162-161), 6 pp.

Springs, J. D. 1957. "Hypochlorination for slime control." Power, 101, 102-104.

Talbot, L. E. et al. 1956. "Treatment of water for cooling purposes." Am. Railway Eng. Assn. Bull., 58, 405-414.

Varma, M. M. and P. W. Reid. 1964. "Comparison of respiration of biological slimes using radiophosphorus." J. Water Poll. Contr. Fed., 36, 176-200.

Wilson, J. N., R. A. Wagner, G. L. Toombs, and A. E. Beecher, Jr. 1960. "Methods for the determination of slimes in rivers." J. Water Pollution Control Fed., 32, 82-88.

Zajic, J. E. 1970. Microbial Biogeochemistry. Academic, New York.

Zobell, C. 1939. "Biological approach to preparation of antifouling paint." Natl. Paint, Varnish and Lacquer Association Circular No. 588 p. 149.

Chapter 14

EUTROPHICATION - PHOSPHORUS AND NITROGEN CONTROL

Literally, eutrophication means "well nourished," whereas in fact it means over-nourished in terms of certain nutritive components. The excess of organic carbon, inorganic nitrogen, or phosphate (also other minerals) causes algae, bacteria, and other flora and fauna to reproduce at very rapid rates. The biosystem that can grow at the most rapid rate usually produces copious quantities of slime and microbial tissue described as a bloom. This bloom indicates an upset in the chemical composition of the stream or lake.

One of the chief indicators of river and lake eutrophication is the algae. This is mainly because they respond to inorganic materials as well as light, their primary energy source. A comparison is made in Table 1 of the plankton in oligotrophic and eutrophic lakes. A reduced number of algae is found in eutrophic lakes, with the quantity of a few species greatly enriched (Rawson, 1956). Eutrophication is accompanied by an increase in both the population of Cyanophyceae and Diatomaceae.

Evidence for eutrophication is often difficult to assess. Davis (1964) has made an evaluation for Lake Erie, Dickman (1959) for Marion Lake in British Columbia, and Lind and Cottam (1969) for Lake Mendota in Wisconsin.

The pollutant chemical (s) causes a shift in the biosystem in the lake, producing an anomolous condition. All habitats in nature have a typical plant and animal micro- and macropopulation. The relationship of this biopopulation

TABLE 1

PLANKTON OF OLIGOTROPHIC AND EUTROPHIC LAKES

	Oligotrophic	Eutrophic
Quantity	Poor	Rich
Variety	Many species	Few species
Distribution	To great depths	Trophogenic layer thin
Diurnal migration	Extensive	Limited
Water-blooms	Very rare	Frequent
Characteristic algal groups and genera	Chlorophyceae (Desmids if Ca-) Staurastrum or Diatomaceae Tabellaria Cyclotella Chrysophyceae Dinobryon	Cyanophyceae Anabaena Aphanizomenon Microcystis and Diatomaceae Melosira Fragilaria Stephanodiscus Asterionella

to its environment is described as ecology. Most biopopulations are quite complex and they are highly dependent on the presence of foods or energy materials. One way of defining and describing these complex ecological groups of life is in terms of food or energy sources as shown in Fig. 1 (Klein, 1962).

A widely accepted concept in ecology is Liebig's Law of Limiting Factor (Odum, 1959). This law states that the rate of growth in activity of some processes in an organism is controlled by some limiting environmental factor. If the limiting factor is removed another factor becomes limiting, etc. Since the environment is continually varying or fluctuating, the influence of this law is easily seen. The law can also be extended to include factors which are inhibiting or toxic to organisms.

Although one factor may limit growth, it would be quite unusual to find a system completely controlled by a single factor. The ecosystem normally responds to variations in temperature, pressure, pH, oxygen concentration, Eh, nitrogen source, carbon source, phosphate source, etc. Thus, ideally it would be better to perform a multivariate analysis of the simultaneous influence of these factors on growth and function. Growth of organisms on non-

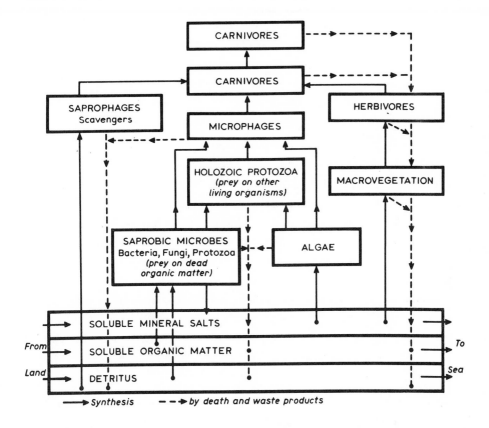

FIG. 1.   Theoretical food cycle classification in a stream bed community (after Klein, 1962).

living materials falls in a field described as saprobiology. This field is often subdivided into ecosystems (Sládecek, 1967) associated with surface waters (limosaprobity), with subsurface waters (kathasaprobity), with sewage and industrial waste (eusaprobity) with velocity of flow (transsaprobity). Important algae in sewage treatment ponds are described by Haughey (1969). Notes on the more important fresh water algae are reviewed by Whitford and Schumacher (1968).

Biological productivity of water can be too high (eutrophication) or too low (oligotrophication). Oligotrophy can be caused by low nutrients or by toxic chemicals. Normally, eutrophication is caused by an excess food or energy source in a water body, part of which may be contributed by photosynthesis.

Mathematical models of productivity are reviewed by O'Connor and Patten (1968). This work touches upon Volterra's classical work on predator-prey populations (1926, 1931), the grazing aspects of phytoplankton limitation (Fleming, 1939), Sverdrup et al. (1942) and the concept of compensation, Ruby's first model of productivity (1946), and Cushing's novel models (1954) in diatom production in the North Sea. These developments and that by Kozlovsky (1968) are most important.

In order to trace the flow of energy in a body of water from solar radiation to a given product four evaluations are normally required (Ivlev, 1966). First, a quantitative determination is made of the formation of primary organic matter, i.e. the fixation of solar energy by plankton and macrophytes. Second, the paths of energy transformation that lead to a specified product must be identified. To do this one must concentrate on the predominate energy conversion steps and ignore secondary reactions. Third, the ecotrophic coefficient must be determined for each step in the food pyramid that leads from primary organic matter up to the product. The term ecotrophic coefficient is usually defined on a dynamic basis as the ratio of a "bioconsumers" intake of a particular food or organism to the production of a product over a rather long period of time, usually a year.

The energy content of the production of each food present multiplied by the corresponding dynamic ecotrophic coefficient defines the energy level and type of material ingested by the organism during the time unit chosen. The sum of the quantities for all foods eaten by consumer organisms gives the total energy within a food pyramid per unit time. To be useful a dynamic ecotrophic coefficient must be carried out on real bodies of water (not in the laboratory). The static ecotrophic coefficient is the ratio of a quantity of food consumed to the total supply available at a given time. The ecotrophic coefficient (dynamic or static) is expressed by

$$\epsilon_i = \frac{\sum^{t} r_i}{{}_oB_i + \sum^{t-1} \Delta B_i}$$

where

ϵ_i is the unknown ecotrophic coefficient for a time t

r_i is the daily ration

oB_i is the original biomass of the given food

ΔB_i is the daily increment of the food

If the quantity of food used is divided by the ecotrophic coefficient, a figure is obtained for the biomass that served as the energy base for the performance of the observed energy transfer. Here, the ecotrophic coefficient becomes a measure of the intensity of utilization of food material at each step.

Since some organic food ends up as CO_2 and H_2O and is partly converted into cellular substance, the energy coefficient of growth must be defined. It simply is the fraction of consumed food converted into cellular substance. With this concept of productivity, which can be applied to any food or ecosystem, general understanding and approach can be made in evaluating the basic process of trophy and an approach can be made in evaluating the basic process of trophic ecology.

References to the food chain by ecologists are common (Rich, 1963). Certainly, energy transformations in the flow of organic matter from one state to another or from one ecosystem to another are important. As a result of photosynthesis and solar radiation (λ_{gR}) energy is converted into organic matter or plants and algae, designated as λ_1 or the first energy level (Fig. 2).

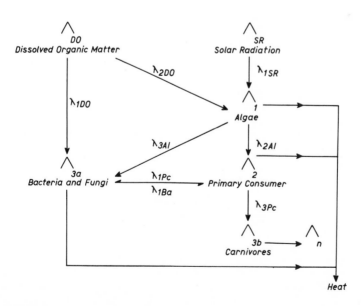

FIG. 2.   Energy flow in an ecosystem (after Saunders, 1963).

This figure shows a first-order system (Saunders, 1963). This is consumed by plant-eating animals or microbes, yielding Λ_2 the second energy level. Λ_2 serves as energy for the carnivores, giving Λ_3 the third energy level. This series progresses to Λ_n with the energy decreasing throughout the progression. At any given stage Λ_n two processes are involved, accession of energy from the previous stage (λ_{n-1}) and release of energy to the next level (λ_{n+1}). The rate of change of the stock of energy at a given level is expressed by Ivlev (196) as

$$\frac{d\Lambda n}{dt} = \lambda_n - \lambda_n^1$$

where

λ_n is the rate of extraction of energy from the previous level (a positive quantity)

$\lambda_n 1$ is the rate of loss of energy to the next level (a negative quantity)

Thus, λ_n is always the true productivity of the level n. Daily energy budgets in environmental fluids are reviewed by Verduin (1965). The relation of other environmental factors to the first-order system is shown in Fig. 3, which describes a second-order system (Saunders, 1963).

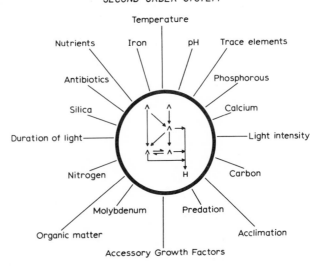

SECOND-ORDER SYSTEM

FIG. 3. A second-order system. Schematic diagram of environmental factors operating to produce the first-order system (after Saunders, 1963).

If one is solely concerned only with the photosynthetic process, Platt
(1969) has developed a coefficient that measures the contribution of the photosynthetic processes to an optical attenuation coefficient. This coefficient
depends on chlorophyll content and depth.

In a eutrophical situation there is always an imbalance in the energy
chain, giving disparity in the ratio of the accession of energy of two adjacent
levels, i.e.

$$\lambda n / \lambda n-1$$

Thus, in considering the production or eutrophication processes in a body of
water, four features are considered: (a) the process of synthesis of the primary organic matter, (b) the paths of transformation of the latter to final
product, (c) the loss of energy from failure to utilize all the edible product
at each trophic level, and (d) loss of energy in the matter of ingestion.

Under conditions where a ecosystem exhibits a need for a certain element,
introduction of the same into water may increase the intensity of photosynthesis several hundred percent. This has been demonstrated with phosphate and
nitrate under natural conditions in oyster banks. Potassium and inorganic
nitrogen (**Pirson,** 1937; Van Hille, 1938) exert a similar effect on <u>Chlorella</u>.

I. PHOSPHATE ABSORPTION AND EXCRETION

Eutrophication of lakes and streams is of rapidly growing interest to all
concerned with water resources. A suggested method of controlling the very
high rate of eutrophication is the lowering of levels of phosphates and nitrogen
in municipal waste plant effluents. Nitrogen control has not received the same
emphasis as phosphate control in lake eutrophication.

According to Sawyer (1968) there are three reasons why nitrogen removal
has not received this attention:

1. It occurs as NH_4, NO_3, NO_2, and organic nitrogen. This makes its overall removal very difficult.

2. About 30% of the total nitrogen applied as fertilizer is leached out and
these waters cannot be effectively treated.

3.   The blue-green algae can fix gaseous nitrogen from the air if no other
source exists.   On this basis it has been suggested that phosphate control is
a more likely point of attack in applying Liebig's law of the minimum to control unwanted algal growth.

A. Sources of Phosphates

The presence of phosphates in nutrient enriched waters has caused some
observers to label phosphates as the key to controlling eutrophication. This
may be too simple an explanation, as Wuhrmann (1968) suggests that the actual
enrichment for rapid growth may well be organic matter from untreated human
sewage. Kuentzal (1969) supports Wuhrmann's view. Even with a broader
view of lake eutrophication, since life cannot exist without phosphate, the control may center around phosphates.

A paper presented by Weaver (1968) indicates (Table 2) the sources of
phosphates entering waters in the U.S. A.

Table 2 indicates that municipal, agricultural, and forest lands are each
responsible for 30% of the total known sources of phosphates, with detergents
contributing about 60% of the municipal phosphate effluent or 18% of the total
known sources. Table 1 assumes that municipal plants have removed 30% of
the phosphate in their influent.

All of the phosphates in detergents are present as polyphosphate and/or
in metaphosphate. However, the bacteria of an activated sludge should hydrolyze these forms to the orthophosphate with an efficiency of about 90%   (Shannon and Lee, 1966).

B. Effective Levels of Phosphate Necessary to
Encourage Excessive Growths of Algae

Connell (1966) indicated that a stream containing 0.3 mg/liter PO_4 appears
to be in the critical tolerance range above which excessive growth will occur
and below which growth will be normal.

C. Removal of Phosphates

Phosphates can be removed from sewage plant effluents by chemical or
biological means. Both approaches seek to change soluble or colloidal PO_4
into insoluble recoverable phosphates.

TABLE 2

ESTIMATE OF PHOSPHORUS CONTRIBUTIONS

FROM VARIOUS SOURCES[a]

Source	Million pounds per year (as P) in U.S.A.	
	Range	Average
Natural		
Rainfall direct	2 - 17	9.0
Aquatic plants	0 - 107	53.0
Birds, fish, fauna	Not known	
Mud	Not known	
Run-off		
Forest land	243 - 587	415
Other land	Not known	
Man Made		
Domestic		
Human and food	137 - 166	150
Washing wastes	250 - 280	265
Industrial		
Food processing	Not known	
Other	Not known	
Run-off		
Urban land	19	19
Cultivated land	110 - 380	245
Feed lot and land with animals	170	170

[a] After Ferguson (1968).

The chemical processes have received the most attention but appear to be more expensive. Table 3 (editorial, 1968; Ferguson, 1968) indicates that the cost of biological treatment may be 5¢/1000 gal compared with about 8¢ for chemical treatment.

TABLE 3

SELECTED OPERATIONAL PHOSPHATE REMOVAL

PROCESS EXPERIENCE IN THE U. S.

	Location	Status	Size in 10^6 gal/ day	Process	Actual or projected cost at 10-20 x 10^6 gal/day (cents per 1000 gal)	Actual or projected phosphate removal efficiency
Chem-ical	Lake Tahoe, Calif.	Start up Feb. 1968	4 to 7.5	Lime ter-tiary with se-paration beds	< 9 cents	> 95% (0.1-1.0 ppm in effluent)
	Las Vegas, Nev.	Opera-tional since 1961 (2 sites)	4	Lime tertiary with single clari-fier	< 4	> 95 (0.5 ppm in efflu-ent)
	Lake Tahoe, Calif.	Opera-tional	2.5-4	Alum tertiary with se-paration beds	< 9	> 95 (0.1-1.0 ppm in effluent)
	Nassau County, Long Island, N. Y.	Opera-tional since 1965	0.6	Alum tertiary with se-paration beds	< 7	> 60
	Lansdale, Pa.	Opera-tional	0.3	Alum tertiary with se-paration beds	< 10	> 90

TABLE 3 (Con'd)

Chemical

Washington, D.C.	Operational	Pilot plant (0.1)	Lime tertiary with dual clarifiers and lime recovery	< 4	> 95

Chemical-biological

Wayne, Mich.	Operational for short period	45	Alkaline waste calcium chloride to raw sewage	--	> 80 (jar test)

Biological

Black River Plant, Baltimore, Md.	Operational (test)	20	Activated sludge phosphate uptake	< 1	> 90
South Central Region	Prior operations (5 sites)	Pilot plant	Activated sludge phosphate uptake	< 1	to 87
Irvine, Calif.	Operational	Pilot plant	Activated sludge phosphate uptake acid elutriation chemical removal	< 5	~ 90

Not all individuals believe that phosphate is the sole controlling compound in lake eutrophication. Kuentzel (1969) states that theory and calculation show that the free CO_2 in most waters lies between 0.4-1.0 mg/liter. In some lakes an algal bloom during a single day would require 110 mg/liter of CO_2. The lake involved was Sebasticook in Maine, which has an alkalinity of 40 (Mackenthum, 1968). Thus, in this instance CO_2 appeared to control the rate of eutrophication, and even though it was limiting, at least in terms of photosynthesis, eutrophication followed. Earlier, Lange (1967) emphasized the effect of carbohydrates on the symbiotic growth of planktonic algae with bacteria. Further analysis has shown that a mixed culture of bacteria and algae grow quite well together, with algae fixing CO_2 and producing O_2 and the bacteria utilizing the O_2 for oxidizing low levels of organic matters present in the water. This type of activity is believed to occur even when phosphate is far more limiting than 0.01 ppm, which was suggested by Sawyer (1947) to be the controlling factor in eutrophication in 17 lakes located in Wisconsin.

These and other data indicate that CO_2 limits algal growth during lake eutrophication and that available phosphate in many lakes would limit the growth of algae, bacteria, and fungi, yet this is not observed. Such facts emphasize that:

1. Many algae grow heterotrophically as well as photosynthetically (Zajic and Chiu, 1969).

2. Many algae grow at faster rates under mesotrophic conditions than they do as autotrophs or heterotrophs (Zajic and Chiu, 1970).

3. Symbiotic growth of algae, bacteria, and fungi has not been investigated adequately.

4. Knowledge of phosphate availability and turnover from lake water, bottom sediments, and organophosphates is inadequate.

5. Lake organic matter is contributing more than was originally suspected to the eutrophication process.

Further, this dilemma points out the need for establishing minimal and maximal concentrations for several factors that control growth rates in biosystems, including limits for carbon, nitrogen, oxygen, phosphate, and potassium. Cognizance should also be taken of the fact that limits for any growth factor vital for growth will limit reproduction; however, a level limiting one member may not be limiting for another. Thus, phosphate can still be considered a prime contributor to eutrophication, and its control will remain important.

Vollenweider (1968) reported data on phosphates in the Great Lakes and on phosphate levels in 20 lakes throughout the world. The degree of phosphate contamination in these lakes was made on the basis of annual phosphate loading,

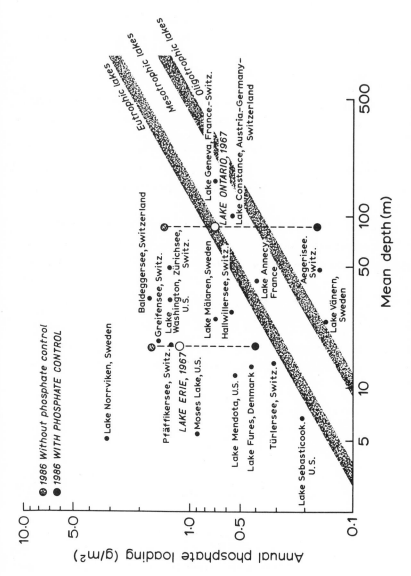

FIG. 4. Phosphate classification of twenty lakes of the world (see editorial, 1969).

mean lake depth, and the degree of enrichment, i.e., oligotrophy, mesotrophy, and eutrophy (Fig. 4). Lake Ontario is mesotrophic and without phosphate control it will be eutrophic by 1986. Effective control would return it to an oligotrophic lake. Lake Erie is already eutrophic by this classification scheme and even with effective control it would only be approaching a mesotrophic condition by 1986. It is doubtful that Lake Erie will ever improve beyond a mesotrophic condition.

1. Biological Phosphate Removal

Environmental engineers have not resolved the problem as to how much phosphate can be removed by the conventional activated sludge treatment.

Jenkins and Menar (1967) have summarized the experience of most operators of the activated sludge process with regard to phosphate removal. They indicate that with the conventional system the maximum limit is about 30% removal.

Levin and Shapiro (1965), Shapiro et al. (1967), and Hennessey et al. (1968) have indicated that the biological uptake can be increased considerably by stripping the phosphate from the return activated sludge and by carrying out more than normal aeration in the aerator. This process appears to indicate a 90% removal of PO_4 at an influent level of 50 mg/liter PO_4.

Figure 5 indicates the process flowsheet as proposed by Hennessey et al (1968). The parameters for the control of this process have been proposed by Backer et al. (1967) as well as by Levin et al. (1965) and Shapiro et al. (1967).

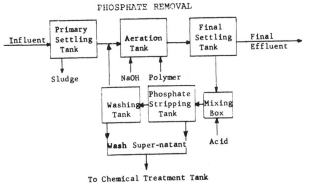

FIG. 5. Biological uptake of phosphate in an activated sludge plant is increased considerably by stripping the phosphate from the return activated sludge (Hennessey, Maki, and Young, 1968).

The organisms which absorb the phosphate are, according to Vacker et al. (1967), <u>Zooglea</u> <u>ramigera</u>, <u>Escherichia</u> <u>intermedium</u>, <u>Bacillus</u> <u>cereus</u>, and <u>Flavobacterium</u> and <u>Pseudomonas</u>. Some protozoa were also involved, e.g., <u>Vorticella</u> and <u>Epistylis</u>.

A generalized procedure of biophosphate removal should include the following which are shown diagrammtically in Fig. 6 which shows a pilot model

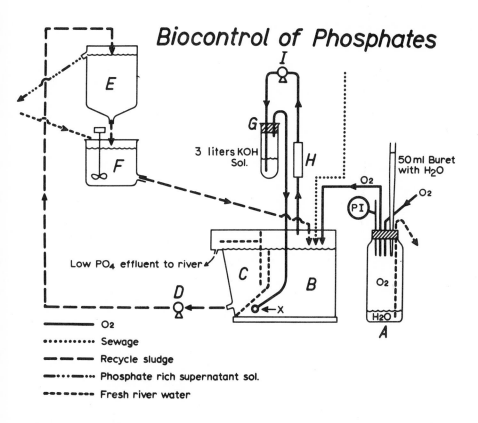

FIG. 6. Flow diagram for pilot model biophosphate removal unit: (A) oxygen reservoir maintained at atmospheric pressure, (3.0 liter); (B) aerator section of activated sludge unit, (35.0 liter); (C) settler section of activated sludge unit, (10.0 liter); (D) pump for return sludge, (3 liter/hr); (E) phosphate stripping tank, (12 liter), (F) mixing tank (activated sludge mixed with river water), (4 liter); (G) KOH sol. to absorb CO_2, (4 liter); (H) flowmeter for air to aerator, (6.0 liter/min), (I) pump for air to aerator, (6.0 liter/min).

unit used in pilot plant studies. Aeration of this unit is with pure oxygen; thus, the gas system is closed and the O_2 needed is added through a makeup line. Flows should be traced for O_2, sewage feed, recycle sludge, phosphate rich effluent, fresh river water, and effluent to river low in phosphate.

1. Oxygen reservoir supplying oxygen for aerobic phase of cycle

2. In the aerator the control conditions are:
   a. Retention time: 4 hr
   b. Suspended solids: 2500–4000 mg/liter
   c. BOD level of influent: 150 mg/liter
   d. Dissolved oxygen: 2 mg/liter in the mid-section to 5 mg/liter at the effluent end
   e. pH: 8.0
   f. To aid settling in the settling tank, Hennessey (1968) proposes addition of a flocculating agent
   g. An air flow rate of about 3700 ft^3/min/10^6 gal of sewage per day is required
   h. Temperature: preferably above 15°C

3. Final settling tank control conditions
   a. A pre-requisite of settling must be to prevent the sludge from becoming anaerobic. This can be achieved by short retention time, or froth flotation or forced aeration to keep the solution aerobic.
   b. The concentration of the returned sludge should be above 6000 mg/liter

4. Pump and mixing tank
   Shapiro (1967) has shown that low pH ~ 5.0 leads to faster release of phosphate. Mixing is required to reduce the pH to this level.

5. Phosphate stripping
   The necessity here is to have anaerobic conditions and sufficient settling time to separate the sludge from the phosphate-rich supernatant solution. Chemical treatment of the phosphate-rich water can be achieved by precipitation with alum and/or lime.

6. Washing tank
   Here the suspended solids with only about 15% of the original solution remaining are washed with river water, again settled, and fed back to the aerator.

7. Alkali is used to scavenge CO_2 from the gas mixture as it is recycled.

8. Flowmeter and gauge for controlling gas flow.

9. Recycle pump for air mixture.

Phosphate removal: Providing there is a sufficiently high concentration of O_2 in solution good absorption of phosphate will take place (Shapiro, 1967; Levin et al., 1966).

According to Levin et al. (1966), the maximum rates of absorption occurred with an hourly oxygen consumption of 100 ml/liter of mixed liquor. The retention time for maximum absorption is about 4 hr.

The higher the suspended solids the better was the phosphate absorption up to about 4000 mg per liter (Levin et al., 1966).

Release of phosphate: After the aerator the sludge is allowed to settle but without allowing the D.O. of the solution to drop below 0.5 p.p.m. (Shapiro et al., 1967).

The returned sludge is removed and pumped to the phosphate stripper. In Fig. 7 a correlation is made (Shapiro et al., 1967) between D.O. levels, Redox potential, and PO_4 release.

Surprisingly, the phosphate release is more a function of Redox potential than D.O. values. Another point to remember is the very rapid decrease of O_2 values once the air supply is cut off.

A correlation is also made between release of PO_4 and temperature of the solution in the stripper (Shapiro, 1967) as shown in Fig. 8 (a). It is apparent that going from $10°$ to $30°C$ can increase the rate seven-fold. However, if the sludge had been acclimatized to lower temperatures, there is a good possibility that the release of phosphate at lower temperatures would approximate those observed at $30°C$.

Figure 8 (b) (Shapiro, 1967) demonstrates the reversibility phosphate release that occurs by supplying air to the solution after an anaerobic condition has occurred.

PHOSPHATE RELEASE

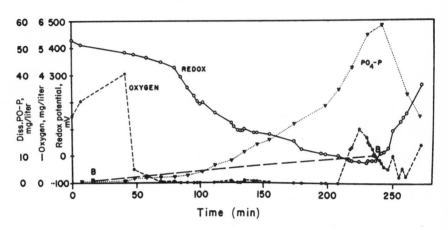

FIG. 7. Release of phosphate from settled sludge as a function of Redox potential and dissolved oxygen. The points B show the small amount of phosphate release in a similar sample under continuous aeration (after Shapiro et al., 1967).

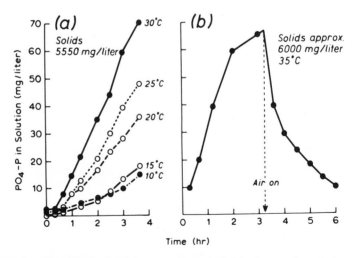

FIG. 8. The effect of (a) temperature and (b) air on phosphate removal and excretion by sludge (after Shapiro, 1967).

TABLE 4

COMPARISON OF ALTERNATIVE PHOSPHORUS REMOVAL METHODS

Process	Removal efficiency %	Type of wastes to be disposed of	Remarks
Conventional biological treatment	10-30	Sludge	
Modified activated sludge	60-80	Sludge	
Chemical precipitation	88-95	Sludge	Removal efficiency and cost depend on chemical used
Chemical precipitation with filtration	95-98	Liquid and sludge	
Sorption	90-98	Liquid and solids	Cost based on water treatment costs
Ion exchange	86-98	Liquid	
Electrochemical treatment	81-85	Liquid and sludge	
Electrodialysis	30-50 [a]	Liquid	
Reverse osmosis	65-95	Liquid	
Distillation	90-98+	Liquid	
Land application	60-90+	None	Large land area required

[a] Removal efficiency per single stage.

Over eleven unit processes have been evaluated for phosphorous removal. Removal efficiences are shown in Table 4 (Eliassen et al., 1968). Over 90% removal was accomplished by chemical precipitation, sorption, ion exchange, reverse osmosis, distillation, and land application. Electrochemical treatment may prove to be the most economical process although not the most efficient.

II. NITROGEN REMOVAL FROM WASTES

Like phosphates, organic and inorganic sources of nitrogen contribute to eutrophication and to pollution. Important compounds of nitrogen are proteins, ammonia nitrogen (NH_3^-), nitrite (NO_2), and nitrate (NO_3). Since many microbial forms fix atmospheric nitrogen, it too can contribute to cause a further imbalance in an eutrophic water body.

If the water body is aerobic, the nitrifying microbes will normally continue to oxidize the reduced forms of nitrogen to nitrate (NO_3). Many heterotrophic microbes, and even some anaerobic heterotrophs and autotrophs, denitrify NO_3 with the production of free nitrogen. Some organic waste or substrate is needed for the heterotrophic population. Methanol can be used as a typical example and the reactions involved are shown here:

$$6H^+ + 6NO_3^- + 5CH_3OH \rightarrow 5CO_2 + 3N_2 + 13H_2O$$

Photosynthetic plants such as algae also remove NO_3 from waters:

$$aCO_2 + cNO_3^- + ePO_4^{\simeq} + (c + 3e) \, H + \frac{1}{2} \, (b - c - 3e) \, H_2O \rightarrow$$

$$C_aH_bN_cO_dP_e + (a + \frac{b}{4} + \frac{5c}{4} - \frac{d}{2} - \frac{5e}{4}) \, O_2$$

Eliassen et al. (1968) estimated that 40-90% of the nitrate in certain waters could be removed by this process.

Ammonia stripping could be accomplished by microbes through nitrification. Ammonia exists in H_2O as

$$NH_4^+ \; \rightleftharpoons \; NH_3 + H^+$$

Above pH 10, over 85% ammonia is present and can be liberated by agitation and sparging the system with air. Packed tray towers equipped with forced aeration are used for this purpose.

Ion exchange may be used for NO_3 removal. Zeolite has been used, e.g. Hector clinoptolite. Anionic resins are used in which common salt can be used to regenerate the resin. Reverse osmosis can be used for both nitrogen and phosphorous containing compounds.

Ammonia is also removed by chlorination:

$$2NH_3 + 2Cl_2 \longrightarrow 2NH_2Cl + 2HCl$$

$$NH_2Cl + Cl_2 \longrightarrow NHCl_2 + HCl$$

$$NH_2Cl + NHCl_2 \longrightarrow N_2 + 3HCl$$

About 6.3 mg/liter of chlorine is needed per mg/liter of ammonia. Best results are obtained between pH 7-9 at $9°C$ with a reaction time of 2 hr.

Removal efficiencies of nitrogen utilizing various procedures are reported in Table 5 (Eliassen et al. 1968). Best efficiencies (over 90%) were obtained with ammonia stripping, denitrification, algae harvesting, ion exchange, reverse osmosis, and distillation. Removal costs were cheapest using ammonia stripping and electrochemical treatment.

TABLE 5

COMPARISON OF ALTERNATIVE NITROGEN REMOVAL METHODS

Process	Removal Efficiency %	Type of wastes to be disposed of	Remarks
Ammonia stripping	80–98	—	Removal efficiency is based on ammonia nitrogen only
Conventional biological treatment	30–50	Sludge	
Anaerobic denitrification	60–95	None	
Algae harvesting	50–90	Liquid and sludge	Large land area required
Ion exchange	80–92	Liquid	Efficiency and cost depend on degree of pretreatment, coag., filt., etc.
Electrochemical treatment	80–85	Liquid and sludge	
Electrodialysis	30–50[a]	Liquid	
Reverse Osmosis	65–95	Liquid	
Distillation	90–98	Liquid	
Land application	–[b]	None	Large land area required

[a] Removal efficiency per single stage.

[b] Removal efficiency depends on form of nitrogen.

REFERENCES

Eutrophication:

Cushing, D. H. 1959. "On the nature of production in the sea." Min. Agric. Fish and Food, U.K. Fish Investig. 22, 1-40.

Davis, C.C. 1964. "Evidence for the eutrophication of Lake Erie from phytoplankton records." Limnology and Oceanography 9, 275-283.

Dickman, M. 1969. "Some effects of lake renewal on phytoplankton productivity and species composition." Limnology and Oceanography 14, 660-665.

Fleming, R. H. 1939. "The control of diatoms populations by grazing." J. Cons. Explor. 14, 210-227.

Haughey, A. 1969. "Further planktonic algae of Auckland sewage treatment ponds and other waters." New Zealand J. Marine and Freshwater Res. 3, 245-261.

Ivlev, V.S. 1966. "Biological productivity." J. Fish. Res. Bd. Can. 23, 1727-1759.

Klein, L. 1962. River Pollution: Causes and Effects. Butterworth, London.

Kozlovsky, D.G. 1968. "A critical evaluation of the trophic level concept." Ecology 49, 48-60.

Lind, C.T. and G. Cottam. 1969. "The submerged aquatics of University Bay: a study in eutrophication." Am. Midland Naturalist 81, 353-369.

O'Connor, J.S. and B.C. Patten. 1968. "Mathematical models of plankton productivity." Reservoir Fisheries Res. Sym., Athens, Georgia, April 5-7, 1967.

Odum, E.P. 1959. Fundamentals of Ecology. Saunders, Philadelphia.

Pirson, A. 1937. Z. Botan. 31.

Platt, T. 1969. "The concept of energy efficiency in primary production." Limnology and Oceanography 1, 18-25.

Rawson, D.S. 1956. "Algal indicators of trophic lake types." Limnology and Oceanography 1, 18-25.

Rich, L.G. 1963. Unit Processes of Sanitary Engineering. Wiley, New York.

Riley, G. A. 1946. "Factors controlling phytoplankton populations in Georges Banks." J. Mar. Res. 6, 54-73.

Saunders, G. W. 1963. "The biological characteristics of fresh water." Proc. 6th Confr. on Great Lakes. Res. Ann Arbor, Mich. June 13-15.

Sládecek, V. 1967. "The ecological and physiological trends in the sapro-biology." Hydrobiologia Acta Hydrobiology, Hydrographica et al., Protisto-log. 30, 513-526.

Sverdrup, H. V. , M. W. Johnson, and R. H. Fleming. 1942. The Oceans. Prentice Hall, Englewood Cliffs, New Jersey.

van Hille, J. 1938. Rec. Trav. Botan. Niedrl 35.

Verduin, J. 1965. "Daily energy budgets in environmental fluids." Am. Biology Teacher 27, 363-370.

Volterra, V. 1926. "Variazioni e fluttuazioni del numero d'individui in specie animali conviventi." Nim. Reale Acad. Nazl. Lincei, Clase Sci. Fis. Nat e Nat. , Ser. Sesta 2, 31-112.

Volterra, V. 1931. "Lecons sur la theorie mathematique de la Lutte pour la Vie." Cahiers Scientifiques (7), Gauthier-Villars, Paris.

Whitford, L. A. and G. J. Schumacher. 1968. "Notes on the ecology of some species of fresh-water algae, "Hydrobiologia Acta Hydrobioloty, Hydro-graphia et al., Protistology 32, 225-236.

Phosphate Control:

Anonymous. 1968. "Phosphate removal processes prove practical."
Env. Sci. Tech. 2, 182.

Brock, T.D. 1966. Principles of Microbiol Ecology, Prentice Hall,
Englewood Cliffs, New Jersey.

Connell, C.H. 1966. "Phosphates in Texas waters - Part "O" of the
1966 Progress Report." The Environmental Health Lab., Dept. of Preventive
Medicine and Public Health, University of Texas Medical Branch.

Editorial. 1969. "Rx phosphates of Great Lakes." Env. Sci. and Tech.
3, 1248-1245.

Eliassen, R., M. Pastner, E. Pastner, and G. Tchobanoglous. 1968.
"Removal of nitrogen and phosphorous." 23rd Purdue Ind. Waste Conf., Purdue Univ. Lafayette, Indiana, pp. 1-17.

Ferguson, A. 1968. "A non-myopic approach to the problems of excess
algae growths." Env. Sci. Tech. 2, 188.

Hennessey, T. L., K. V. Maki and E. Y. Young. 1968. "Phosphorus
removal in waste water by a modified activated sludge process." Paper presented at the Fed. Water Poll. Contr. Assoc. sponsored workshop in PO_4
removal. Chicago, Ill.

Jenkins, D. and A. Menar. 1967. "The fate of phosphorus in sewage
treatment processes 1." Prim. Sed. and Act. Sludge, S.E.R.L. Report # 67-6,
University of California.

Kuentzel, L. E. 1969. "Bacteria, carbon dioxide, and algae blooms."
J. Water Pollution Contr. Fed. 41, 1737-1747.

Lange, W. 1967. "Effect of carbohydrates on the symbolic growth of
planktonic blue-green algae with bacteria." Nature 215, 1277-1278.

Levin, G. V. and D. G. Shaheen. 1966. "Metabolic removal of phosphate
from sewage effluent." Presented before the Am. Chem. Soc., Division of
Microbiol. Chem and Tech., N. Y., September.

Levin, G. V. and J. Shapiro. 1965. "Metabolic uptake of PO_4 by waste
water organisms." J. Water Pollution Contr. Fed. 39, 11, 1810.

Mackenthum, K. M., L.E. Keup and R. K. Stewart. 1968. "Nutrients and algae in Lake Sebasticook, Maine." J. Water Pollution Contr. Fed. 40, 72-81.

Mesarovic, M.D. 1968. Systems Theory and Biology. Springer-Verlag, New York.

Phillipson, J. 1966. Ecological Energetics, Edward Arnold, London.

Sawyer, C. N. 1968. "The need for nutrient control." J. Water Pollution Contr. Fed. 40, 363.

Sawyer, C. N. 1947. "Fertilization of lakes by agricultural and urban drainage." New England Water Works Assoc. 61, 109-127.

Shannon, J. E., and G. F. Lee. 1966. "Hydrolysis of condensed phosphates in natural waters." Air and Water Res. Int. J. 10, 735-756.

Shapiro, J. 1967. "Induced rapid release and uptake of phosphate by microorganisms." Science 155, 1269.

Shapiro, J., G. V. Levin and H. G. Zea. 1967. "Anoxically-induced PO_4 release in waste water treatment," J. Water Pollution Contr. Fed. 39, 11, 1810-1818.

Vacker, D., C. H. Connell and W. N. Wells. 1967. "Phosphate removal through municipal waste water treatment" at San Antonio, Texas, J. Water Pollution Contr. Fed. 39, 5,750.

Vollenweider, R. A. 1968. "Eutrophication of water by phosphorous." Rev. Ital. Sost. Grasse, 45, 99-107.

Weaver, P. J. 1968. "Phosphates in surface waters and detergents." J. Water Pollution Contr. Fed. Annual Conference Issue, September.

Wuhrmann, K. 1968. "Objectives, technology and results of phosphate and nitrogen removal processes," in Advances in Water Quality Improvements. Ed. Gloyne, E. F. and W. W. Eckenfelder, Jr., University of Texas Press.

Zajic, J. E. and Y. S. Chiu. 1970. " Heterotrophic growth of algae" in Properties and Products of Algae. Plenum Press, New York.

Chapter 15

ODOR AND TASTE

Taste, odor, and color are the criteria that the general public use in determining whether or not a water is "pure." If the tap water is unpalatable,
the public will resort to drinking untreated spring or well water that is more
palatable but more prone to be unsafe from a health viewpoint. Thus, good
quality potable water should be free of any objectionable tastes and odors
when it is used for cooking, drinking, and bathing purposes. Complete removal is not always necessary, but taste- and odor-producing substances
must be reduced to concentrations that are below detectable levels.

According to Sigworth (1957) and Hainer et al. (1954), the problem of
water palatability is one of odor rather than taste. There are only four true
taste sensations: sour, sweet, salty, and bitter. Methods used in determining flavor profiles are reported by Caul et al. (1957). All other taste
sensations commonly ascribed to the sense of taste are actually odors, even
though the sensation is not noticed until the material is taken into the mouth.
All particular tastes arise from one or a combination of the aforementioned

taste sensations (Adams, 1951) along with the odor or odors associated with the substance. Therefore, "odor" is a separate sense but "taste" is a combination of taste and odor. Odor is often rated on a sensory scale of 1 to 4 (Table 1), which can be correlated with concentration (Fig. 1).

TABLE 1

SENSORY SCALE FOR EVALUATING ODOR INTENSITY LEVELS

Degree of odor intensity	Description
0	Odorless
1/2	Perception threshold
1	Recognition threshold
2	Definite
3	Strong
4	Overpowering

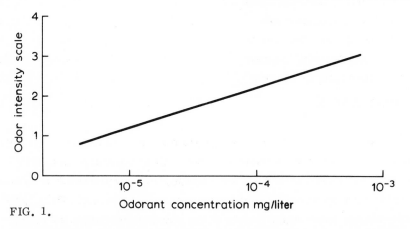

FIG. 1.

Objectionable tastes and odors occur in potable water supplies because of the presence of foreign substances, usually organic, such as algae and decaying vegetation (Sproul and Ryckman, 1961). The smallest dose giving a response is known as the minimum identifiable odor (MIO). Some inorganics, such as hydrogen sulfide, also cause objectionable odor.

Algae are the most frequent cause of odors in water supplies, with decaying vegetation a close second. Trade wastes, although they occur less frequently, can also be troublesome.

Alga types most frequently causing odor problems are: Anabaena, Aphanizomenon, Asterionella, Diatoma, Dinobryon, Synedra, Synura, and Tabellaria.

Tastes and odors due to dissolved minerals from underground strata do not constitute a problem except in certain localized areas. Surface waters may have an odor problem. Chlorinous odors due to an excess of residual chlorine could be a serious issue; however, chlorine is lost rapidly in all open bodies of water.

The odorous substances are usually present in potable water supplies in very small quantities (a few micrograms per liter) and are often very complex. It is therefore impractical and often impossible to isolate and identify positively the odor-producing chemicals. Evaluation of odors is thus dependent on the sense of smell. The standard test method is given by the American Public Health Association, ASTM (1964), and the Manual on Sensory Testing Methods (1968). These describe the threshold dilution method (TDM) for odor evaluation and are based on the principle that odor perception is a log function of the concentration (Weber-Fechner law); this method is described on p. 343. Major groups of odorous compounds are summarized in Table 2. Mercaptans, amines, sulfides, and aldehydes are the big offenders. Minimal identifiable odors in water for common odoriferous groups are shown in Table 3.

TABLE 2

MAJOR GROUPS OF ODOROUS COMPOUNDS[a]

Compounds
Mercaptans
Amines
Indoles
Organic acids
Skatoles
Alcohols
Aldehydes
Ketones (e.g., heptanone)
Sulfides
Simple compounds of nitrogens, bromine, chlorine, and sulfur

[a] Middleton and Rosien, 1956.

TABLE 3

MINIMAL IDENTIFIABLE ODOR OF SOME
COMMON ODOROUS COMPOUNDS IN WATER

Compound	Formula	Concentration, mg/liter
Ammonia	NH_3	0.037
Sulfur dioxide	SO_2	0.009
Hydrogen cyanide	HCN	0.001
Hydrogen sulfide	H_2S	0.00018
Iso-valeric acid	$CH_3(CH_2)_3COOH$	0.000005
Ethyl mercaptan	CH_3CH_2SH	0.00000066

Corrective treatment of objectionable odors in potable water supply can
be achieved with various degrees of success by the following methods:

1. Adsorption with activated carbon.
2. Chemical oxidation with
   a. Chlorine dioxide (ClO_2)
   b. Superchlorination (with subsequent declorination)
   c. Free residual chlorination
   d. Ozone
3. Aeration

According to Sigworth (1957), reports of completely successful corrective treatments show that activated carbon is the most superior, with an
overall effectiveness of 86% compared with 25% for chlorine dioxide, 23% for
superclorination with subsequent dechlorination, 17% for free residual chlorination, and 9% for aeration. Data on the use of ozone are not available.

Subsurface waters generally have no offensive odors although some regions of the country have water tainted to varying degrees with hydrogen sulfide. The odor in soils is imported by Actinomycetes (Henley et al., 1969).
A compound identified as 5-methyl-3-heptanone imparts this odor.

Surface waters often have taste and odor problems, particularly in
regions of dense population. These are generally caused by one or more of
the following: industrial or domestic pollution, algae, aquatic vegetation,
decaying vegetation (Baker, 1961), and runoff from agricultural regions.

There is only one rapid odor-measuring test (Sproul and Ryckman, 1961), which in most instances in plant practice is too awkward and only qualitative. The method is to heat a sample of the water in question to about 60°C (the temperature of most domestic hot water heaters), and if it has a noticeable odor then some form of taste and odor treatment should be employed.

I.   TREATMENT

Taste and odor control can be divided into categories of physical and chemical treatment (Howard, 1946).   Physical treatment is the adsorption of the odorous material onto solid medium. Alum floc, for example, will adsorb some odorous material; however, it is generally not sufficiently effective to be considered a form of taste and odor treatment. Adsorption onto activated carbon is the only form of physical treatment employed in the water industry. The reaction between certain forms of activated carbon and many organic compounds has been intensively studied and it has been proved that carbon will effectively remove extremely minute amounts of organics from water. The reaction proceeds to equilibrium quite rapidly. It is then a fairly simple matter to separate the carbon from the water.

There are two general methods of using activated carbon: granular carbon in columns or beds and powdered carbon intimately mixed with the water. In the former case a relatively small quantity of water contacts a large amount of carbon for a short time, while with powdered carbon a large quantity of water contacts a small amount of carbon for a relatively long time. Originally, the powdered and granular carbons differed only in physical size; however, certain forms of carbon are available now only in powdered form because of restricted manufacturing processes. Consequently, there is available a wider selection of powdered forms than granular ones.

Generally, granular carbon treatment consists of loading large plastic-lined pressure filters with carbon through which all clarified water is passed. The plastic lining is necessary because of the seriousness of electrolytic corrosion between the walls of the pressure vessel and the carbon granules. Units are expensive to build, require a large inventory of carbon, and can only be operated economically if the spent carbon is regenerated. This usually requires an expensive regenerating furnace. Also, multiple filters are required. Occasionally, for small treatment plants with intermittent taste problems, it is considered practicable to accept the high capital cost in view of the tremendous capacity and ease of automatic operation. The filters are normally expected to last three to five years before being replaced.

Powdered carbon is added anywhere in the treatment plant that will allow sufficient contact time. Generally, it is fed into the raw water line ahead of the flocculators in quantities, determined by jar tests, and at levels suffici- ent to remove the taste. In all carbon treatment, prechlorination may im- prove efficiency but normally should be avoided since it will sometimes "fix" the tastes and make them more difficult to remove (Hyndshaw, 1962). Pow- dered carbon allows greater efficiency, relatively inexpensive feeding equipment, and a lower inventory of chemical stocks. It is difficult to handle and creates a potential explosion hazard.

The theory of chemical treatment is quite simple, i.e., oxidize the or- ganic contaminants using powerful oxidizing agents. The most commonly used of these (based largely on economics) are potassium permanganate, chlorine dioxide, chlorine, calcium or sodium hypochlorite, air, and ozone. The chlorine-containing compounds have the added advantage of providing residual protection required by health authorities.

A. Potassium Permanganate ($KMnO_4$)

This chemical has come into wide use only comparatively recently with a decrease in cost (Cherry, 1962). Although expensive when compared with chlorine, it is very cheap on the basis of taste improvement per dollar since very little is required. $KMnO_4$ earned an undeserved bad reputation from early experiences in treating water supplies, because it has a deep purple color and its reaction products are dark brown. Understandably, if either of these colored compounds got into the distribution system, the consumer would be upset. Actually there is no good reason for the bad reputation. The purple color and other colored reaction products are insoluble compounds that can be removed on a filter. This chemical can be fed with inexpensive equipment and control is quite simple once the $KMnO_4$ demand has been es- tablished. This is completed by treating samples of water with various doses of $KMnO_4$ and determining visually at 30-min time intervals what dosage has been consumed. Any change in the organic loading of the raw water will result in an increase or decrease in the permanganate demand.

Benefits from the use of $KMnO_4$ are the generally sharp decrease in the post-chlorine demand and resultant economies from reduced expenditures for chlorine.

Ozone (O_3) is the most powerful oxidant available to the water treatment industry. It has been known and used in water treatment in Europe for over 60 years but has never been popular in North America. This is probably due to the very expensive auxiliary equipment necessary. Ozone must be generated on-site since its great reactivity precludes easy transportation and storage. In ozone generation, air (or oxygen) is dried to a dew point of $-60^{\circ}C$ and then passed between dielectric surfaces across which is maintained a very high alternating voltage. A silent electric discharge converts some of the oxygen in the gas to ozone. This gas mixture is then contacted with the water either in an injector diffuser, or tray tower. The drying process is essential because ozone generation decreases sharply in the presence of moisture. Water is removed by refrigeration and/or dessicators. If the organic turbidity is too high (in algae) pretreatment may be necessary to remove solids. The high reactivity of ozone means that very little ozone is required (generally less than 2 ppm) and contact time is short (less than 5 min). For some unexplained reason the O_3/air ratio in the feed gas has a strong effect on the efficiency of the treatment, while above a certain minimum dose the O_3/water ratio is not too important. Ozone is an excellent disinfectant, but since it gives no residual protection it does not have approval in many states.

Chlorine dioxide (ClO_2) is a pungent yellow gas generated (at least in the water industry) by reacting sodium chlorite $(NaClO_3)$ and hypochlorous acid (HOCl) at pH 4 or less in a packed tower. This compound is reasonably soluble in water and possesses 2-1/2 times the oxidizing power of chlorine. Its cost, however, is many times greater than chlorine and therefore its use is rather limited, mainly being confined to the treatment of phenol for which it is specific. In practice, the ClO_2 generator is operated with sufficient excess HOCl to meet the basic chlorine demand while limiting the $NaClO_3$ feed to generate sufficient ClO_2 to oxidize the phenols.

Chlorine (Cl_2) is never truly fed as the elemental liquid; rather it is dissolved in water to form hypochlorous acid or it is sold as the sodium or calcium salt of this acid, both of which liberate OCl^- ion (the active agent) in

water. This compound is the most commonly used oxidant in the water treatment industry (William, 1952; Riddick, 1951), largely because of its low cost
and ready availability. Chlorine is not the best taste- and odor-treatment
chemical because it often enhances odors (Harlock and Dowlin, 1958).
More commonly it is a disinfectant that renders the water potable. For disinfection purposes it is sufficient that the chlorine combines with organic
compounds (primarily chloramines), which then provide the disinfecting
action.

To remove taste and odor, sufficient excess chlorine must be supplied
to substitute all available organic matter plus an excess, which then oxidizes
the compounds to odorless end-products. The excess chlorine must then be
removed. This process is termed "free residual chlorination," although
break-point and superchlorination are terms sometimes used. The process
works extremely well. In highly polluted waters the chlorine demand is
great and cost of treatment is high; dechlorination presents an added cost.
In some instances, problems of nitrogen trichloride formation occur in the
distribution system. This is due to chlorine reacting with albuminoid nitrogen, a slow reaction requiring several hours to complete.

Safe operation of this system almost demands an automatic chlorine
residual controller operating a sulfur dioxide feeder and/or an ammonia
feeder to provide chloramines for distribution system protection. This
equipment is expensive and requires routine maintenance.

Air is the least effective oxidant for controlling taste and odor; however,
it does have practical use in taste and odor control, particularly in oxidizing
ferrous ion to ferric in iron-bearing subterranean supplies. Iron imparts
a bitter, somewhat metallic taste to water when present in quantities exceeding about 3 ppm. Also air will remove hydrogen sulfide. The hydrogen
sulfide-bearing waters are aerated to remove excess carbon dioxide and oxidize hydrogen sulfide to elemental sulfur or to SO_4^{2-}. The reaction reaches
equilibrium, however, at about 3 ppm H_2S and pH 6-7. The remaining H_2S is
then removed by oxidizing with chlorine and filtering to remove the sulfur.
The air is introduced by injectors, diffusers, packed towers, or nozzles.

II. DETERMINATION OF THE THRESHOLD ODOR NUMBER
BY THE THRESHOLD DILUTION PROCEDURE

Various methods have been proposed to determine the odor intensity of
water and waste waters (Fox and Gex, 1957; Hemeon, 1968; Mills et al.,
1963). The procedure summarized here is based on the Standard Method for
the Examination of Water, approved by the American Public Health Association, the American Water Works Association, and the Water Pollution Control
Federation (1965). This method is believed to be practical and economical
in terms of time and personnel (Benforado et al., 1968) and to be adequate
for most odor problems encountered in potable water treatment plants; however, it may not be applicable in industry (Byrd, 1964; Ettinger and Middleton,
1956).

A. General Precautions

The persons selected to make odor tests should be normally sensitive to
odors. Panels of five persons or more are recommended to overcome the
deficiences of using one tester, although useful data can be obtained by one
tester only (Turk, 1967).

A calibrating standard is required when it is necessary to depend on one
tester or to compare the sensitivity of different individuals. Dilutions of
n-butyl alcohol in the range 0.125-4 mg/liter are used frequently for calibra-
tion. However, a calibrating standard is not required when several observers
are testing the same sample at the same location.

Smoking and eating should be avoided just prior to making the test.
Scented soaps, lotions, and perfumes should be avoided. The test room
should be free from distractions, draft, and odor. A special odor-free room
is desirable. Tests should not be prolonged to the point of fatigue. Someone
other than those in the test group should code and prepare the dilutions.

Temperatures of the samples and the blanks should be kept within $\pm 1^{\circ}C$
of the desired temperature of the test. Threshold values vary with temperature. Tests are normally made at $40^{\circ}C$. At times, however, odor tests are
made at $60^{\circ}C$ to detect odors that might be missed at lower temperatures.
Therefore, the temperature of the test should always be reported. Odor tests
made at $60^{\circ}C$ are referred to as "Hot Threshold Tests." "Cold Threshold
Tests" are carried out at $40^{\circ}C$.

Important: sniffing of the testing flasks should always be started with the most dilute sample. Test rooms should be of specific design and construction (Deininger and McKinley, 1954).

B. Description of Odor

It is useful to describe first the nature of the odor of the sample water. This description could help in determining the source of the odor and in selecting the proper treatment method. Shake 200 ml of sample in a 500-ml glass-stoppered Erlenmeyer flask that has been previously heated to 40 or 60°C. Sniff the odor lightly and record the nature of the odor in terms that best describe the odor: aromatic, chemical, disagreeable, earthy, moldy, etc.

C. Determination of Threshold Odor Number

Apparatus required: (a) constant temperature bath ($40\pm1^{\circ}$C); (b) sample bottles: glass-stoppered BOD bottles; (c) odor flasks: 500 ml, wide-mouth, glass-stoppered Erlenmeyer type; (d) thermometer: $0-110^{\circ}$C, chemical or metal stem dial type; (e) pipets: 200, 100, 50, 25, 10 ml, graduated in tenths; (f) odor-free water bottles: 2 liter, glass stoppered; (g) odor-free water generator.

Chemicals required: (a) odor-free water; (b) activated carbon, water purification grade; (c) n-butyl alcohol.

1. Preparation of Odor-free Water

Odor-free water should be prepared by passing tap water through the generator at a rate of 0.1 liter/min, or through a glass column 3 ft long and 2 in. in diameter, packed with granular activated carbon, at a flow rate of less than 11 liters/hr. Glass connections and tubing should be used in making the apparatus. The column ends may be packed with glass wool to support the column.

When the generator is first started, water should be passed through for a short while to wash out the carbon fines before collecting the odor-free water needed for the test. The activated carbon should be removed after treating approximately 20 liters of tap water, or more often as necessary. The life of the carbon will vary with condition and amount of water filtered.

Detection of odor or residual chlorine in the water coming through indicates that a change of carbon is needed.

Odor-free water should not be stored but should be prepared on the day the test is made.

2. Method

The "Threshold Odor Number (TO)" is the ratio by which the odor-bearing sample has to be diluted with odor-free water for the odor to be just detectable.

The total volume of sample and odor-free water used in each test is 200 ml. The proper volume of odor-free water should be put in the testing flask first. The sample is then pipeted into the odor-free water.

(a) Determine first the approximate range of TO by adding 200, 50, 12, and 2.8 ml of sample water to separate 500 ml (Table 4) Erlenmeyer flasks containing odor-free water to make a total volume of 200 ml. A blank containing odor-free water only should also be prepared. Heat the dilutions and the flask to $40^{\circ}C$.

TABLE 4

DILUTIONS FOR VARIOUS ODOR INTENSITIES

Sample volume in which odor was first detected	200 ml	50 ml	12 ml	2.8 ml
Volume (ml) of sample to be diluted to 200 ml	200	50	12	2.8
	140	35	8.3	2.0
	100	25	5.7	1.4
	70	17	4.0	1.0
	50	12	2.8	

(b) Shake the blank, remove the stopper, and sniff the vapors. Then test the most dilute sample in the same way. Proceed to higher concentrations until odor is clearly detected.

(c) On the basis of the preliminary results, prepare another set of dilutions using Table 4 as a guide.

Two blanks should be inserted in the series. Dilutions and blanks should be heated to $40^{\circ}C$. The test panel should then shake and smell each flask in sequence, beginning with the least concentrated sample and comparing with the blanks until odor is detected with certainty.

(d) Observations should be recorded by indicating with a + or - sign whether odor is detected for each testing flask, as the following sample:

ml sample diluted to 200 ml →12 0 17 0 25 0 35 0 50

response → - - - - + - + - +

TO

3. Calculation of TO

As mentioned previously, the TO is the dilution ratio at which odor is just detectable. It is calculated as follows:

$$TO = \frac{ml \text{ of sample} + ml \text{ odor-free water}}{ml \text{ of sample}}$$

If a total volume of 200 ml was used, the values shown in Table 5 give the TO corresponding to common dilutions.

NOTE: If the sample requires additional dilution, prepare an intermediate dilution with 20 ml of sample diluted to 200 ml. Proceed as in (a), (b), and (c) and multiply the TO by 10 to correct for the intermediate dilution.

TABLE 5

THRESHOLD ODOR NUMBERS CORRESPONDING TO
VARIOUS DILUTIONS

Milliliter sample diluted to 200 ml	Threshold odor number	Milliliter sample diluted to 200 ml	Threshold odor number
200	1	12	17
140	1.4	8.3	24
100	2	5.7	35
70	3	4	50
50	4	2.8	70
35	6	2	100
25	8	1.4	140
17	12	1.0	200

Odor thresholds have been reported for over 53 important odorous chemicals (Leonardos et al., 1968). All concentrations are calculated as ppm by volume of air. The minimum threshold observed was with trimethylamine at

0.00021 ppm (0.21 ppb). Methylene chloride does not have a distinct smell at a concentration of 215 ppm. The 53 compounds studied represent a range of six orders of magnitude in threshold concentrations (Table 6). Odor descriptions (other than chemical name) used to describe the odor quality of the odorant chemical are included in Table 6.

III. ODOR THRESHOLD VALUES

A. Sulfur Containing Compounds

As a group, compounds with the sulfur atom in their structure have the lowest thresholds of the compounds evaluated (Table 7). All the sulfides with the exception of carbon disulfide and sulfur dioxide have threshold concentrations at the ppb level.

B. Nitrogen Compounds

Values for amines, nitrates, etc., are given in Table 8.

C. Oxygenated Compounds

Table 9 lists the odor thresholds of oxygenated compounds according to chemical class. Extent of oxidation of the ethanol series (ethanol, acetaldehyde, acetic acid) does not appear to have an effect on the odor threshold trend. One might expect a lower threshold as the oxidation state is increased. Chloral (trichlorinated analog of acetaldehyde) does have a substantially lower threshold (0.047 ppm) than acetaldehyde. Considering other chlorine-containing compounds studied, it is not possible to make a generalization as to the effect of chlorination on the odor threshold.

D. Unsaturated Compounds

The presence of unsaturation in an odorant chemical is not associated with low threshold concentrations (Table 10). Ethyl acrylate and methyl methacrylate are isomeric; however, the threshold concentrations are quite different. The odor descriptions do differ (see Table 6). This disparity points up the difficulty of making extended generalizations pertaining to the odor threshold based on similar chemical structures.

TABLE 6

ODOR THRESHOLDS IN AIR (ppm VOLUME)

Chemical	Odor Threshold	Odor Description
Acetaldehyde	0.21	Green sweet, oxidized
Acetic acid	1.0	Sour
Acetone	100.0	Chemical sweet, pungent
Acrolein	0.21	Burnt, sweet, pungent
Acrylonitrile	21.4	Onion/garlic-pungency
Allyl chloride	0.47	Garlic-onion pungency, green
Amine, dimethyl	0.047	Fishy
Amine, monomethyl	0.021	Fishy, pungent
Amine, trimethyl	0.00021	Fishy, pungent
Ammonia	46.8	Barn-like, pungent
Aniline	1.0	Oily, solvent, pungent
Benzene	4.68	Solvent
Benzyl chloride	0.047	Solvent
Benzyl sulfide	0.0021	Cedary, sulfidy
Bromine	0.047	Irritation, bleach
Butyric acid	0.001	Sour
Carbon disulfide	0.21	Vegetable sulfide
Carbon tetrachloride (chlorination of CS_2)	21.4	Sweet, pungent, feeling factor
Carbon tetrachloride (chlorination of CH_4)	100.0	
Chloral	0.047	Sweet, fruity
Chlorine	0.314	Pungent, bleach
Dimethylacetamide	46.8	Amine, burnt, oily
Dimethylformamide	100.0	Fishy, floral, pungent
Dimethyl sulfide	0.001	Cooked vegetable
Diphenyl ether (perfume grade)	0.1	
Diphenyl sulfide	0.0047	

Ethanol (synthetic)	10.0	Sweet, floral
Ethyl acrylate	0.00047	Hot plastic, earthy
Ethyl mercaptan	0.001	Earthy, sulfidy
Formaldehyde	1.0	Hay/straw-like, pungent
Hydrochloric acid gas	10.0	Pungent, burnt
Hydrogen sulfide (from Na_2S)	0.0047	Boiled eggs
Hydrogen sulfide gas	0.00047	
Methanol	100.0	Sweet, fruity
Methyl chloride	(Above 10 ppm)	
Methylene chloride	214.0	
Methyl ethyl ketone	10.0	Sweet
Methyl isobutyl ketone	0.47	Sweet, floral
Methyl mercaptan	0.0021	Sulfidy, pungent
Methyl methacrylate	0.21	Pungent, sulfidy
Monochlorobenzene	0.21	Chlorinated, moth balls
Nitrobenzene	0.0047	Shoe polish, pungent
Paracresol	0.001	Tar-like, pungent
Paraxylene	0.47	Sweet, moth balls
Perchloroethylene	4.68	Chlorinated
Phenol	0.047	Medicinal, sweet
Phosgene	1.0	Hay-like
Phosphine	0.021	Oniony, mustard
Pyridine	0.021	Burnt, pungent, diamine
Styrene (inhibited)	0.1	Solventy, rubbery
Styrene (uninhibited)	0.047	Solventy, rubbery, plasticy
Sulfur dichloride	0.001	Sulfidy
Sulfur dioxide	0.47	Oppressive
Toluene (from coke)	4.68	Floral, pungent, solventy
Toluene (from petroleum)	2.14	Moth balls, rubbery
Tolylene diisocyanate	2.14	Medicated bandage, pungent
Trichloroethylene	21.4	Solventy

TABLE 7

ODOR THRESHOLDS - SULFUR COMPOUNDS (ppm VOLUME)

Hydrogen sulfide (cylinder)	0.00047
Dimethyl sulfide	0.0010
Ethyl mercaptan	0.0010
Sulfur dichloride	0.0010
Benzyl sulfide	0.0021
Methyl mercaptan	0.0021
Hydrogen sulfide (Na_2S)	0.0047
Diphenyl sulfide	0.0047
Carbon disulfide	0.21
Sulfur dioxide	0.47

TABLE 8

ODOR THRESHOLDS - NITROGENOUS COMPOUNDS (ppm VOLUME)

Trimethyl amine	0.00021
Nitrobenzene	0.0047
Monomethyl amine	0.021
Pyridine	0.021
Dimethyl amine	0.047
Aniline	1.0
Acrylonitrile	21.4
Ammonia	46.8
Dimethyl acetamide	46.8
Dimethyl formamide	100.0

E.   Benzenoid Compounds

The effect of substitution on the benzene ring also produces a wide variation in odor thresholds (Table 11). A single methylation of the benzene ring (toluene) lowers the thresholds one concentration step. Further methylation (p-xylene) succeeds in lowering the odor threshold by an order of magnitude. The $-CH_2Cl$ group (as in benzyl chloride) produces an odor threshold that is 100 times lower than benzene. The only generalization possible with benzenoid-type materials is that substitution on the ring reduces the odor threshold by as much as a thousandfold depending on the nature of the group added. It

TABLE 9

ODOR THRESHOLDS - OXYGENATED COMPOUNDS (ppm VOLUME)

Carbonyls

Chloral	0.047
Acetaldehyde	0.21
Acrolein	0.21
Methyl isobutyl ketone	0.47
Phosgene	1.0
Formaldehyde	1.0
Methyl ethyl ketone	10.0
Acetone	100.0

Esters

Ethyl acrylate	0.00047
Methyl methacrylate	0.21

Carboxylic acids

Butyric acid	0.001
Acetic acid	1.0

Alcohols

p-Cresol	0.001
Phenol	0.047
Ethanol	10.0
Methanol	100.0

It should be noted that in addition to the threshold concentration for a given chemical compound there are at least three sensory attributes and two psychological parameters to be considered in evaluating the compound's contribution to odor problems. The sensory attributes are (a) the change in intensity (odor strength) with concentration, (b) the qualitative character of the odor, and (c) the type and degree of interaction with other odorants.

Psychologically, people react either positively or negatively to odor types - they like them or dislike them. A problem is usually associated with those that people dislike. In addition, people react more strongly to things that are different and, therefore, what is normal or expected is frequently accepted, be it good or bad.

TABLE 10

ODOR THRESHOLDS – VARIOUSLY SUBSTITUTED ETHYLENIC
COMPOUNDS (ppm VOLUME)

Chemical	R group (s)	Odor thresholds
Ethyl acrylate	$-C:OOC_2H_5$	0.0047
Styrene	$-C_6H_5$	0.047
Acrolein	$-HC:O$	0.21
Methyl methacrylate	$-CH_3$, $-C:OOCH_3$	0.21
Allyl chloride	$-CH_2Cl$	0.47
Tetrachloroethylene	$-Cl$ (4X)	4.68
Trichloroethylene	$-Cl$ (3X)	21.4
Acrylonitrile	$-C \equiv N$	21.4

TABLE 11

ODOR THRESHOLDS – VARIOUSLY SUBSTITUTED BENZENOID
COMPOUNDS (ppm VOLUME)

Chemical	R group (s)	Odor thresholds
Paracresol	$-OH$, $-CH_3$	0.0010
Nitrobenzene	$-NO_2$	0.0047
Phenol	$-OH$	0.0470
Benzyl chloride	$-CH_2Cl$	0.0470
Styrene	$-CH:CH_2$	0.0470
Monochlorobenzene	$-Cl$	0.21
p-Xylene	$-CH_3$, $-CH_3$	0.47
Aniline	$-NH_2$	1.0
Toluene	$-CH_3$	2.14
Benzene	$-H$	4.68

REFERENCES

Adams, E. M. 1951. Air pollution abatement manual. Physical Effects Gasoline pp. 22-26 Chap. 5., Manual sheet p-6. Manufacturing Chem. Assoc., Washington, D. C.

American Standard Testing Methods (ASTM) Standards. 1964. 23, 267-274.

Baker, A. 1961. "Tast and odor studies in water." J. Water Pollution County Fed. 33, 1066-1099.

Benforado, D. M., W. J. Rotella, and D. L. Horton. 1968. "Development of an odor panel for evaluation of odor control equipment." 61st Annual Meetg. Air Pollution Control Assn. June 23-27, St. Paul, Minn.

Byrd, J. F., H. A. Mills, C. H. Schellhase, and H. E. Stokes 1964. Solving a major odor problem in a chemical process. J. Air Pollution Control Assn. 14, 509-516.

Caul, J. F., S. E. Cairncross and L. B. Sjostrom. 1968. "The flavor profile in review." Perfumery Essent. Oil Record 49, 130-133.

Cherry, A. K. 1962. "Use of potassium permanganate in water." J. Am. Water Works Assn. 54, 417-424.

Deininger, N. and R. W. Mckinley. 1954. "The design, construction and use of an odor test room." ASTM Special Tech. Publ. No. 164.

Ettinger, M. B. and F. M. Middleton, 1956. "Plant facilities and human factors in taste and odor control." J. Am. Water Works Assn. 48, 1265-1273.

Fox, E. A. and V. E. Gex. 1957. "Procedure for measuring odor concentration in air and gases." J. Air Pollution Control Assn. 7, 60-61.

Hainer, R. M., A. G. Emslie, and A. Jacobson. 1954. "Theory of olfaction." Ann. New York Acad. Sci. 58, 158-173.

Harlock, R. and R. Dowlin. 1958. "Use of chlorine for control of odors caused by algae." J. Am. Water Works Assn. 50, 29-32.

Hemeon, W. C. L. 1968. "Technique and apparatus for quantitative measurement of odor emissions." J. Air Pollution Control Assn. 18. 166-170.

Henley, D. E., W. H. Glaze, and J. K. G. Silvey. 1969. "Isolation and identification and an odor compound produced by a selected aquatic actinomycete." Env. Sci. Tech. 3, 268-271.

Howard, N. J. 1946. "Progress in the treatment of water for taste and odor correction." Water and Sewage 84, 14-15, 54.

Hyndshaw, A. Y. 1962. "Treatment application points for activated carbon." J. Am. Water Works Assn. 54, 91-98.

Leonardos, G., D. A. Kendall and N. J. Barnard. 1968. "Odor threshold determinations of 53 odorant chemicals." 61st Annual Mteeg. Air Pollution Control Assn. June 23-27, St. Paul, Minn.

Manual on Sensory Testing Methods. 1968. ASTM Special Tech. Pub. No. 434, May.

Middleton, F. M. and A. A. Rosen. 1956. "Organic contaminants affecting the quality of water." Public Health Rpts. (U.S.) 71, 1125-1233.

Mills, J. L., T. Walsh, D. Luedtke and K. Smith. 1963. "Quantitative odor measurement." J. Air Pollution Control Assn., 13, 467-475.

Riddick, T. M. 1951. "Controlling taste, odor, and colour with free residual chlorination." J. Am. Water Works Assn. 43, 545-552.

Sigworth, E. A. 1957. "Control of odor and taste in water supplies." J. Am. Water Works Assn. 49, 1507-1521.

Sproul, O. J. and D. W. Ryckman. 1961. "The significance of trace organics in water pollution." J. Water Pollution Centr. Fed. 33, 1188-1198.

Standard Methods for the Examination of Water and Waste Water, 1965. 12th Ed. APHA, AWWA, WPCF, p. 304-311.

Turk, 1967. Selection and Training of Judges for Sensory Evaluation of the Intensity and Character of Diesel Exhaust Odors, U.S. Dept. of Health, Education & Welfare Public Health Service, Bureau of Disease Prevention and Environmental Control, Cincinnati, Ohio.

Williams, D. B. 1952. "How to solve odor problems in water-chlorination practice. Water and Sewage Works 99, 358-364.

Chapter 16

COLOR REMOVAL

The color of water (F.W.P.C.A., 1968) is attributed to substances in solution after removal of suspensoids. Color may be of organic or mineral origin. Organic sources are humic materials, peat, plankton, rooted and floating aquatic plants, tannins, etc. Inorganic sources are metallic substances such as iron and manganese compounds, clays, chemicals, and dyes. Many industries discharge materials that contribute to the color of water. Among them are pulp and paper mills, textile mills, refineries, manufacturers of chemicals, dyes, and explosives, mines, and tanneries.

I. COLOR CRITERIA

After centrifugation of a water sample, true color is determined by the platinum-cobalt method (A.P.H.A., A.W.W.A., W.P.C.F., 1965). The standard Hazen unit of color is the color produced by 1 mg/liter of platinum in the form of the chloroplatinate $(PtCl_6)^{2+}$ ion. Color in excess of 30 Hazen units may limit photosynthesis and have a deleterious effect upon aquatic life, particularly phytoplankton, and benthic organisms. The Saskatchewan Water Resources Commission (S.W.R.C., 1968) and the Environmental Health Services Division of Alberta (E.H.S.D., 1968) both have adopted a criterion of 30 Hazen units as being acceptable for surface waters used for recreation and propagation of fish and wildlife.

II. REMOVAL OF COLOR

Because color also causes an undesirable esthetic effect on water used for domestic supplies and is detrimental for various industrial processes, public opinion is the prime motivating force for removal or reducing of color in waste waters.

Advanced waste treatment processes considered for removal of dissolved organics resistant to biological treatment include distillation, freezing, reverse osmosis, and adsorption on activated carbon, both granular and powdered. Adsorption on granular carbon is the most advanced of these processes and is particularly amenable for the removal of color (Cooper and Hager, 1966).

A. Manufacturing of Activated Carbon

Source materials include coal, wood, sawdust, peat, lignite, pulp mill char, and sewage sludge. In the first step of manufacturing the raw material is carbonized in the absence of air at 600°C to remove the bulk of volatile materials (F.W.P.C.A., 1968). Activation is associated with controlled oxidation at elevated temperatures. Activation gases are usually steam or carbon dioxide and the temperature is held between 800 and 1000°C. The time of activation varies between 30 min and 24 hr, depending on the oxidizing conditions and on the quality of active carbon desired.

B. Principles of Adsorption

Adsorption is a phenomenon by which solutes in a solution are attracted to and adhere to the surface of solid materials. Activated carbon is a particularly good adsorbent because it has an extremely large surface area per unit of volume (areas of 1000 m^2/g are not uncommon).

The driving force for adsorption is a function of concentration of the material to be adsorbed, active surface area available, type of adsorbate and, to a lesser degree, pH and temperature.

One empirical formula, the Freundlich adsorption isotherm, states that the quantity of material adsorbed per unit weight of adsorbent is proportional to the concentration of solute in equilibrium with the adsorbent. This equation can be written as follows:

$$X/M = kC^{1/n}$$

where   x is the unit of material adsorbed

           M is the weight of adsorbent

           C is the equilibrium concentration of materials remaining
              unadsorbed in solution

           k and n are constants that have different values for each solute,
              adsorbent, pH, temperature.

Figure 1 shows isotherm data taken by adding different amounts of adsorbent to given quantities of sewage waste water and measuring residual chemical oxygen demands after equilibrium has been reached.

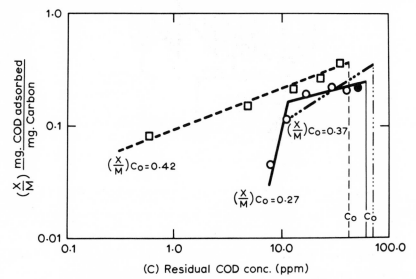

FIG. 1. COD isotherms using virgin carbon and different secondary sewage effluents (after Masse, 1967).

Some generalities relating to the type of materials adsorbed by carbon are given here:

1.   Weak electrolytes are adsorbed better than strong electrolytes.

2.   The more ionic a material, the more difficult it is to adsorb.

3.   Sparingly soluble materials are generally adsorbed better than highly soluble materials.

4.   High molecular-weight materials may be adsorbed better than those of low molecular weight.

Pilot plant studies have been completed on activated carbon units. A four-stage series contact unit was used. A 200 gal/min pilot plant for treating municipal secondary effluent has been operated at Pomona, California.

Figure 2 shows a schematic of this four-stage series contact plant.

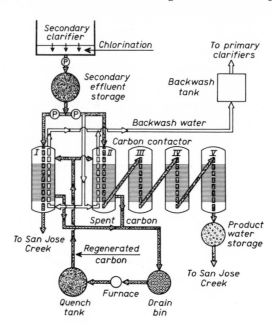

FIG. 2. Schematic flow diagram of a carbon adsorption pilot plant (after Masse, 1967).

Contactor No. 1 is operated at 200 gal/min (7 gal/min/ft^2) and the effluent is discharged to waste. When only 25% of the dissolved COD is being removed, the carbon is removed from the contactor, regenerated, and returned to service.

Contactor Nos. 2 through 5 are operated in series. When the dissolved COD from the last contactor in series reaches 12 ppm, the carbon in the first contactor in series is removed, regenerated thermally, and returned to the same contactor. The piping is rearranged so that the contactor containing the freshly regenerated carbon is last in the series. By this mode of operation, the carbon in the first contactor in series is nearly in equilibrium with the feed (and therefore contains the maximum possible loading of organics) and the newest carbon on line contacts the water just before it leaves the system as a product.

Table 1 gives data on the quality of the feed and product during 14 months of operation. About 80% of the dissolved COD has been removed from the secondary effluent.

TABLE 1

AVERAGE WATER QUALITY CHARACTERISTICS OF MAIN CARBON COLUMN
(Masse, 1967)
JUNE 1965 TO AUGUST 1966

Parameter	Column Influent	Column Effluent
Suspended Solids mg/liter	10	< 1
COD, mg/liter	47	9.5
Dissolved COD, mg/liter	31	7
TOC, mg/liter	13	2.5
Nitrate, as N mg/liter	6.7	3.7
Turbidity, JTU	10.3	1.6
Color	30	3
Odor	12	1
CCE, mg/liter	--	0.014

C. Regeneration

Two types of regeneration have been studied -- chemical and thermal.
Chemical regeneration, even by the strongest oxidants, is uneconomical.
Hydrogen peroxide, the most effective regenerant tried, showed recoveries of
70% of the adsorption capacity on the first regeneration, 50% on the second,
and 20% on the third. In a large plant, a furnace can be used to obtain almost
complete regeneration. Most installations use a multiple-hearth-type furnace
with rabble arms to move the carbon from hearth to hearth.

D. Costs

A breakdown of the total costs for a 10^7 gal/day plant is shown in Table
2. This plant was designed to be identical in flow path and operating conditions
to the Pomona Plant, i.e., four pressure contactors in series, no pretreatment,
average product dissolved COD of 7 ppm. The capital cost is 16.7 cents/gal/day
and the total cost is 8.3 cents/1000 gal. Figure 3 shows the estimated cost data
for plants up to 10^8 gal/day. These estimates may change as larger plants
are built and more cost data are accumulated.

TABLE 2

COST ESTIMATE (Masse, 1967)

FOR 10 MGD ACTIVATED CARBON TREATMENT PLANT

ITEM	COST/10^6 GAL
Capital (1,670,000; 15 yr. at 4%)	$ 41.00
Power (1 cent/kWh)	8.50
Labor (4 men)	15.00
Maintenance	5.00
Carbon Regeneration:	
Power, Gas & Water	2.50
Makeup Carbon (10% Loss)	11.00
Total	$ 83.00

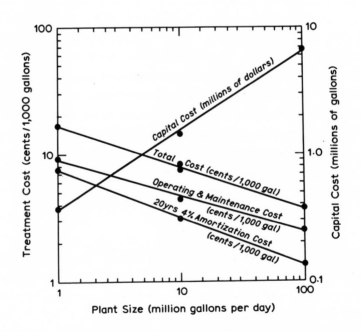

FIG. 3. Cost estimates for granular carbon adsorption (after Masse, 1967).

Activated carbon is effective in reducing color by removing organic pollutants and especially biologically resistant (refractory) compounds from waste waters. Waste treatment systems utilizing regenerable granular activated carbon have been developed which demonstrate the technical and economic feasibility of this approach.

More research is required to determine the cost of equipment for obtaining true countercurrent flows on a large scale. True countercurrent flow of carbon and the waste water will utilize the maximum capacity of the carbon.

Excessive color creates an undesirable effect on fresh water biota as well as an undesirable esthetic effect. Since color control is expensive, public opinion and regulatory agencies will have to provide the impetus for instituting correction.

E. Problem: Using Pulp and Paper Mill Waste

Sulfite waste liquors contain in dissolved or very finely divided suspension approximately half of the weight of the wood used for pulping, and comprise fiber-binding substances such as lignin, pectin, hemicelluloses, sulfites, tannin, and numerous other organic and inorganic substances (McKee and Wolf, 1963). They are usually colored yellow-orange, high in BOD, with a tendency to foam in receiving streams.

Secondary treatment (mechanically aerated lagoons, 4-6 days' retention) may reduce the BOD and suspended solids concentration to nominal levels. More emphasis on color restrictions may be imposed in the future and the following example depicts the terms of references for color control considerations of a typical pulp and paper mill.

A pulp and paper mill discharges 20×10^6 Imp. gal/day of secondary treated waste water (2000 Hazen units) into a river flowing at the rate of 3000 ft^3/sec. The increase in river color is

$$\frac{20 \times 10^6 \times 10 \times 2000}{10^6 \times 3000 \times \frac{5400}{1000}} = 24.6 \text{ Hazen units}$$

Note that this increase is just below the maximum criterion of 30 Hazen units for surface water used for recreation and propagation of fish and wildlife.

From Fig. 3, a preliminary cost estimate for color reduction of the waste water would require a capital expenditure of $2.5 million. The total cost including operation, maintenance, and amortization would be 6.5 cents/ 1000 gal or $475,000/year.

The reduction of color requires individual study into each of the various types of waste waters. One recommendation would be an empirical test in which the waste to be purified is contacted with granular carbon in small scale colums (Fornwalt and Hutchins, 1966).

REFERENCES

A.P.H.A., A.W.W.A., W.P.C.F. 1965. Standard Methods for the Examination of Water and Wastewater, 12th Ed. American Public Health Association, 111-113, 345-349.

Cooper, J. C. and D. G. Hager. 1966. "Water reclamation with activated carbon." Chem. Eng. Progress 62, 85-90.

Environmental Health Services Division, Alberta. 1968. Surface Water Quality Criteria. Saskatchewan Water Resources Commission, 6.

Federal Water Pollution Control Administration. 1968. Water Quality Criteria, Supt. of Documents, U.S. Government Printing Office, 48.

Fornwalt, H. J. and R. A. Hutchins. 1966. "Purifying liquids with activated carbon." Atlas Chem. Ind. Inc., Reprint No. D-101.

Masse, A. N. 1967. Technical Seminar on Advanced Waste Treatment, U. S. Dept. of the Interior, Federal Water Pollution Control Administration, pp. 5-1 to 5-20.

Moysa, R. W. and P. G. Shewchuk. 1967. "Computer program for color analysis." Environmental Health Services Division, Alberta.

McKee, J. E. and H. W. Wolf. 1963. Water Quality Criteria, 2nd Ed. California State Water Quality Board, 277.

Saskatchewan Water Resources Commission. 1968. Water Quality Criteria. Saskatchewan Water Resources Commission, 6.

Chapter 17

BIODETERIORATION

Food, structural materials, or any material of economic importance are subject to chemical and biochemical deterioration. Microbes produce enzymes, which catalyze the degradation of almost an infinite number of compounds, e.g., paper, wood, textiles, grain, steel, concrete, paint, hydrocarbons, etc. Other biological forms also contribute to biodeterioration. Barnacles and molluscs cause fouling of ships and marine structures, rodents destroy grains, insects damage crops, and birds are a hazard to aircraft.

Many factors in the environment are involved in deterioration. For any given material or product a particular set of chemicals and physical conditions are required for maximum stability. The principal elements of the environment influencing biodeterioration are pH, oxidation-reduction potential, air, temperature, and humidity. All of these must be evaluated in relation to the subject material and the biosystem causing degradation. The kind of biodeterioration and the rate are important. Numerous biological agents must be considered.

I. CELLULOSE

Cellulose is a polymer of glucose. The glucose units are connected by $1,4,\beta$- glucosidic bonds and there are more than 3000 glucose residues per molecule. Many bacteria and fungi produce the enzymes cellulase and

363

cellobiase, which degrade cellulose. The thermophilic actinomycetes are
very active in this respect as well as many anaerobic bacteria commonly found
in the rumin of animals (Enebo and Pehrson, 1960).

There are over a dozen fungi that cause timber deterioration. Two are
responsible for 95% of the fungal decay in buildings (Scott, 1968). Meruluis
lacrymans, the dry rot fungus, causes a brown cubical rot. It attacks timber
with a moisture content of 20% or more. Lignin is not attacked. Deep cracks
or splits appear along and across the grain at intervals between 12 to 76 mm.
The other fungus, Coniophora cerebella, which favors wet conditions, is called
the wet rot fungus. Timber must have a moisture content of 25% or more for
it to grow. Splits and cracks usually run with the grain. Other wood rotting
fungi are Poria vaillanti, Paxilluis panuoides, and Lentinus lepideus.

Sap stain in timber is usually caused by fungi but it does not affect the
strength of the wood. Wood is best protected by keeping the moisture level less
than 20% and keeping O_2 away from cellulose surface, since fungi living on
cellulose requires oxygen. Insects such as termites and powder post bettles
also attack wood.

Roaches live as scavengers and feed on a wide variety of organic materials.
Cardboard, gums, and mucilaginous materials are good examples. Larvae of
cloth moths eat all kinds of animal fibers such as raw wood, mohair, fur, hair,
bristles, feathers, and down. Silverfish feed on starchy or cellulose products.

In a marine environment, wooden ships are attacked in salt or brackish
water. Molluscian borers or ship worms attack wharves and other structures,
which often collapse before it is realized they are infected. Crustacean borers
commonly called "gribble" have worldwide distribution but are less spectacular than ship worms in activity. They form interlacing, branching burrows
about 1.2 mm in diameter and seldom extend more than 12 mm below the surface.
Marine borers do not occur when strong currents flow over the surface. All
types of marine organisms foul ships. Some of the groups most frequently
encountered are crustaceans (barnacles), bryozoans (filamentous and encrusta-
ceous), tunicates, annelids (tube worms), algae (sea weeds), coelenterates
(corals, anemone), and mollusks (clams, oysters, mussels). Water velocity
also controls the ability of fouling organisms to attach to underwater structures.

II. PAINTS

The organisms definitely known to cause coatings deterioration include fungi and bacteria, mosses, termintes, marine organisms associated with fouling, such as slime bacteria and protozoa, crustaceans, bryozoans, tunicates, annelids, algae, coelenterates, and mollusks, and the marine boring organisms.

The organisms involved in deterioration of paints are capable of utilizing certain susceptible ingredients of coatings as food. A yeast-like microbe Pullularia pullulans causes severe destruction and black-brown staining of wall paints.

Natural vegetable oils such as linseed, soybean, cottonseed, castor, and many plasticizers containing fatty acid residues all appear to permit growth of the microorganisms. Many other organic compounds are also present, i.e., constituents of water paints such as glue, casein, or other proteinaceous substances, which will support mold growth.

The bacteria are involved in black discoloration of films and may play a part in the decomposition of water paints while still in the container. Desulfovibrio desulfuricans has the ability to reduce sulfates to sulfides, which in the presence of lead pigments form black insoluble salts of lead sulfide.

Attack by fungi and bacteria on vinyl films used for rain-coating material is sufficiently severe to stiffen the coating and render it useless. Here, the fungi and bacteria are believed to attack the susceptible oil-base plasticizer, thus inducing stiffness. The use of resistant plasticizers or the incorporation of fungicides and bactericides in susceptible plasticizers offers solutions to the problem.

Both the organisms that foul metal ships and those that foul or bore into wooden hulls ruin the organic finish during the course of the organisms life processes. Fouling organisms, in the very minute and young stage, attach themselves to ship bottoms while the ship is at rest in fouling waters. The weight represented by the mass of organisms can attain almost unbelievable proportions, depending upon the size of the ship, and may run into a matter of hundreds of tons.

Antifouling paints depend for their action upon the slow leaching out of toxic material, the resulting solution acting as a poison to the marine organisms.

Cuprous oxide is the most widely used for this purpose.

III. METALS

The annual corrosion cost of metals is estimated to cost in the billions
of dollars. The involvement of autotrophic iron microbes in this process has
been reviewed by Zajic (1969). Iron, one of the most frequently used metals
for fabricating purposes, undergoes rapid oxidation and reduction by microbes.
If sulfide or sulfur is available it may be oxidized to the corrosive solvent
sulfuric acid. If sulfate and some oxidized organic compounds are available,
anaerobic conditions may develop and microbes such as Desulfovibrio desulfuricans and Clostridium nigrificans may produce the corrosive gas, H_2S.
The involvement of sulfur in a variety of compounds common to liquids and
gases is best examined by studying the sulfur cycle (Fig. 1).

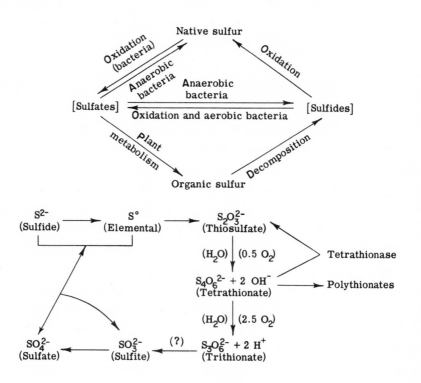

FIG. 1. Bioconversion of sulfur and related compounds (upper figure
shows general cycle; lower shows proposed intermediates formed).

Iron is not the only metal attacked by microbes. Vanadium, chromium, copper, nickel, cobalt, and any metal sulfide are subject to microbial degradation. The involvement of microbes in metal transformations is examined in total in the field of microbial biogeochemistry (Zajic, 1969).

IV. STONE, BRICK, CONCRETE, AND PLASTER

Concrete and like porous materials are attacked by bacteria, algae, and lichen. Damage and type of deterioration can be classified into (a) mechanical processes, (b) chemical (and biochemical) processes (assimilatory and dissimilatory), and (c) functional processes, e.g., soiling.

Freezing and thawing do great mechanical damage. Microbes increase this effect through their water binding capacity. They exert a similar effect when the porous medium undergoes repeated shrinking and relaxation when going through drying and moistening cycles (Jaag, 1945).

Attack by assimilatory chemical processes is not likely to be a problem since the structure material involved is primarily inorganic. Soiling and staining are serious. This is usually caused by fungi, algae, and lichen, which prefer surfaces for growth. Jaag found the alga Gloeocapsa kützingiana growing with fungal hyphae on the surface and in the crevices of stone. Certain of the lichen produce lichenic acids, which chelating properties that solubilize many minerals.

The deterioration on concrete is controlled by pH. At pH 8.4-12.0 carbonation of free lime occurs, at 7.5-8.4 fixation of H_2S is observed, and this is oxidized to thiosulfuric and polythionic acids by Thiobacillus thioparus. Between pH 2.0-7.5, Thiobacillus thiooxidans and Thiobacillus concretivorous oxidize H_2S, S, and S^{2-} to sulfuric acids. The H_2S is produced by sulfate reduction by anaerobic microbes under highly specific conditions (Zajic, 1969).

V. GRAINS

Deterioration of foods and grains far exceeds that of structural materials. Estimates indicate that annual losses of wheat exceed 5% or over 50 billion lb. Other cereal crops show higher losses.

Field fungi invading the kernels of plants are species of Alternaria, Cladosporium, Helminthosporium, and Fusarium. Other fungi readily attack grains during storage: a dozen species of Aspergillus as well as species of Penicillum and Sporendonema. The most common storage fungus is Aspergillus glaucus. The moisture level and temperature influences and controls the development of storage fungi. Below 15.0% moisture Aspergillus restrictus predominates; above 15.0% Aspergillus repens, Aspergillus amstelodami, and Aspergillus ruber predominate. Growth is rapid at temperatures of 30-32°C. Wheat should be dried to a moisture level of 13.0% if it is to be stored for more than two months.

Grain-infesting mites Acarus siro and Tyrophagus costellanic are found in abundance in many commercially stored wheats (Christensen and Kaufmann, 1969). Fungi growing on wheat or stored grain increases the temperature of the grain through their respiratory processes (Gilman and Barron, 1930). This heat often becomes dangerous, with temperatures reaching 55°C or greater.

Certain fungi found in grain produce mycotoxins for man and animal. Burnside et al. in 1957 isolated cultures of Aspergillus flavus and Penicillium rubrum from moldy grain. Addition of these cultures to moist corn fed to swine resulted in death to the animals within a few days. Aflatoxin has since been isolated from Aspergillus flavus. This toxin is found in peanuts that have not been dried under controlled conditions. There are several types of aflatoxin, e.g., aflatoxin M, aflatoxin B-1 with the latter being the most toxic (Ciegler et al. 1966). Aspergillus ochraceus produces a toxin causing liver injury in duckling and mice (Theron et al., 1966). Fusarium tricinctum causes blight and decay to seeds and fruits, producing alimentary toxic aleukia (ATA) affecting 10% of the population in certain areas (Joffe, 1965). A similar fungus, Fusarium graminearum, produces an estrogenic syndrome in swine, causing a severe disturbance of the reproductive organs and processes. These and other cultures of fungi commonly found in grains are toxic to rats.

Data on insects, mites, and rodents that infect grains are well documented by Christiansen and Kaufman (1969).

REFERENCES

Burnside, J. E., W. L. Sippel, J. Forgacs, W. T. Carll, M. B. Atwood and C. R. Doll. 1957. "A disease of swine and cattle caused by eating moldy corn, II: experimental production with pure cultures of molds." Am. J. Vet. Res. 18, 817-824.

17. BIODETERIORATION 369

Christiansen, C. M. and H. H. Kaufmann. 1969. Grain Storage. University of Minn. Press, Minneapolis.

Ciegler, A., R. E. Peterson, A. A. Lagoda and H. H. Hall. 1966. "Aflatoxin production and degradation by Aspergillus flavus in 20 liter fermentors." Appl. Microbiol. 14, 826-833.

Enebo, L. and S. O. Pehrson. 1960. "Thermophilic digestion of a mixture of sewage sludge and cellulosic materials." Acta Polytechnica Scand. (Chem. & Metal Series). 281, 1-36.

Gilman, J. C. and D. H. Barron. 1930. "Effect of molds on temperature of stored grain." Plant Physiology 5, 565-573.

Greathouse, Glenn, A. and Carl, J. Wessel. 1954. Deterioration of Materials: Causes and Preventive Techniques. Reinhold, New York.

Joffe, A. Z. 1965. Toxin Production by Cereal Fungi Causing Toxic Alimentary aleukia in Man, in Mycotoxins, in Foodstuffs, Ed. G. N. Wogan. M.I.T. Press, Cambridge, Mass. pp. 77-85.

Jaag, O. 1945. "Untersuchungen über die Vegetation und Biologie de Algen des Nackten Gesteins in den Alpen, im Jura und im schweizerischen Mittelland," Beiträge Kryptogamenflora Schweiz 9, Heft 3.

Keap, W. M. and S. K. Morrell. 1968. "Microbiological deterioration of rubbers and plastics." J. Appl. Chem. 18.

Rich, L. G. 1963. Unit Processes of Sanitary Engineering. Wiley, New York.

Scott, G. A. 1968. Deterioration and Preservation of Timber in Building. Longmans Green.

Theron, J. J., R.J. van der Merwe, N. Liebenberg, H. H. B. Joubert and W. Nel. 1966. "Acute liver injury in ducklings and rats as a result of ochratoxin poisoning. " J. Path. Bacteriol. 91, 521-529.

Zajic, J. E. 1969. Microbial Biogeochemistry. Academic Press, New York.

Chapter 18

VIRUS DETECTION IN WATER

Before discussing the gauze-pad technique for recovering viruses from sewers, a few general comments concerning viruses are required. An understanding of the nature of viruses progresses only as effective methods of measuring virus concentrations are developed. At first, relative virus concentration was expressed in terms of units of infectivity. In time, procedures were devised to enumerate total virus particles by indirect and direct means, thus permitting studies on the quantitative relationship between the physical particles and their biological activities.

There are two basic reasons for pursuing this relationship. An investigation of the physical and particularly the chemical nature of a virus in its extracellular state requires assurance of the identity of the infectious particle and of the purity of its preparation. Confidence in these areas depends in part upon a knowledge of the ratio of infectivity to total virus particles. Additionally, an appreciation of this ratio is essential to an understanding of the mechanisms of virus replication, which then must be related to the minimum amount of virus necessary to initiate infection.

I. TITRATION OF VIRUS INFECTIVITY

The titration of a virus is essentially a determination of the smallest amount of a virus suspension that will produce some manifestation of disease in a susceptible host. Two types of manifestations are commonly observed, namely, a generalized or systematic infection usually referred to as an all-or-none response, or the production of local lesions. Assays based on the all-or-none response are carried out by innoculating a measured volume of serial (logarithmic) dilutions of the virus suspension into groups of a suitable host species. Estimates are then made on the minimal volume capable of producing infection. This volume is called an infectious unit, and the titre of the original virus suspension is expressed as the number of infectious units per milliliter. Such titres represent the relative infectivity of virus preparations without indicating the number of virus particles per dose. For comparison of titres, therefore, it is necessary to define the conditions of titration with care.

In assay systems where a virus produces local lesions (such as bacteriophage plaques and monolayer cell plaques), serial dilutions are also employed but the titre is expressed as the number of plaque-forming units per milliliter of original virus suspensions. Since theoretically each lesion is initiated by a single virus particle, an estimate can be made of the absolute number of virus particles in the original suspension if the efficiency of lesion production is known.

Conceptions of the potential capacity of the individual virus particle to induce infection are long-standing and were developed initially in studies on bacteriophage. These studies were quite easy to use to determine plaque counting procedures and were extended subsequently to animal and plant viruses.

Before tissue culture procedures became available, the principles of animal virus work differed greatly from those applicable to bacteriophage. Approach to the problem of infectivity measurement of the latter could be made by direct plaque or infectious center studies comparable with methods used at present for tissue cultures. In contrast, investigations with animal viruses involved use of intact higher animals, and efforts were made to interpret the data by statistical analysis.

II. SAMPLING: THE USE OF GAUZE PADS IN THE ISOLATION OF VIRUSES IN SEWERS

Moore (1948) first described the use of gauze pads (now also known as sewer swabs) for the isolation of pathogenic intestinal bacteria from sewage. By placing the swabs in the main and branch sewers in a community and testing the sewage expressed from the swabs for all-or-none response by monkey inoculation it was possible to trace the source of infection back to the house of the carrier. MacCallum et al. (1962) next used gauze pads in the detection of poliomyelitis virus in sewers.

The use of gauze pads is best described by outlining the work of Chin and Gravelle (1961) on the isolation of enterovirus from sewage. A major urban epidemic of poliomyelitis occurred in Des Moines, Iowa, in 1959. One of the phases of the study involved sewage sampling to determine the distribution of poliovirus throughout the community. Eight sewage sampling points were selected in different parts of the city. The sampling points were selected at manholes of branch sewers serving 233 to 1013 dwellings. The estimated population of the area varied from 930 to 3310 persons. The locations of sampling points are shown in Fig. 1, and the characteristics of the populations of the eight areas are summarized in Table 1.

A. Sampling Methods

Sewage samples were collected by suspending gauze pads (Fig. 2) in the sewage flow. The swabs were prepared by covering a cotton-filled maternity pad with tabular surgical stockinette. The ends of the stocking were tied with string. The swabs were suspended through the manhole and immersed in the sewage flow. After three days in the sewage flow, they were pulled up and placed in polyethylene bags. Six drops of 1.0 N sodium hydroxide were added, and the bag was kneaded to express the fluid from the swab. The sewage was transferred to glass jars and adjusted with solidium hydroxide to a final pH of 8.

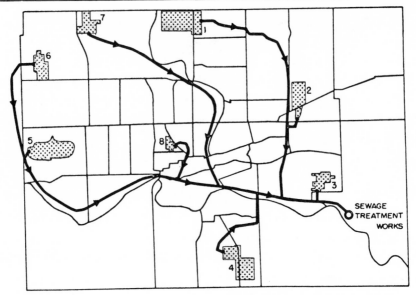

FIG. 1. Locations of the eight sewage sampling areas in Des Moines, Iowa.

B. Preparation of Sewage

Each raw sewage specimen was clarified by preliminary centrifugation at 2000 rpm for 15 min before further processing by each of three methods: (a) unconcentrated method, (b) centrifuge method, and (c) ion-exchange resin method. However the results obtained by these preparations revealed that concentration by centrifugation is superior to the other methods, and this is the only method described herein.

In the centrifuge method, 1 ml of 2.4% gelatin was added to 40 ml of clarified sewage following the method of Baron (1957). The specimen was centrifuged at 39,000 rpm for 1 hr. The supernatant fluid was discarded and the sediment resuspended in 4 ml of phosphate-buffered saline. This treatment yielded approximately a ten-fold concentration of original material. The resuspended pellet was treated with antibiotics and centrifuged at 15,000 rpm for 30 min at 40°C to remove bacteria. The supernatant fluid was stored at -20°C until tested.

Each prepared sewage sample was inoculated into three monkey kidney tissue cultures. The inoculated tissue cultures were observed microscopically every day for cytopathogenic effect (CPE). Each tube culture was harvested by freezing when the CPE exceeded 75%. Samples were considered

TABLE 1

CHARACTERISTICS OF THE EIGHT AREAS FROM WHICH SEWAGE SAMPLES
WERE OBTAINED, IN DES MOINES, IOWA, 1959

Sampling area	Socio-economic classification[a]	No. of family units	Estimated population
1	Upper middle	1013	3110
2	Lower middle	721	2580
3	Lower	261	980
4	Lower middle	233	930
5	Upper	503	1850
6	Upper middle	545	1790
7	Upper	327	1090
8	Lower	337	1200

[a] Based on estimated values of housing units in the area.

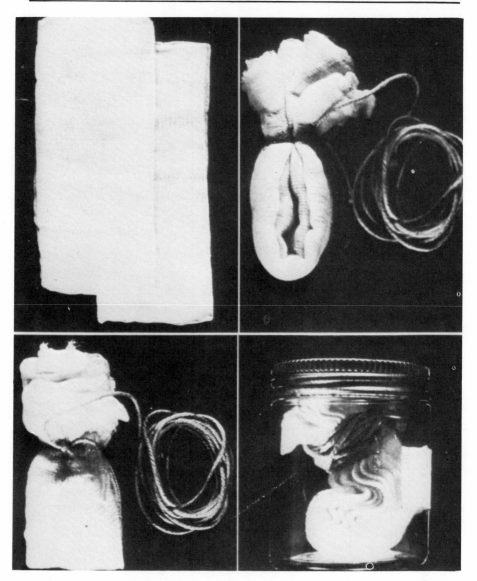

FIG. 2. **Sewer swab.** (Top, left) two pieces of cotton gauze 80 x 40 cm folded to 12 x 6 cm. (Top, right; and bottom, left) swab tied with string. (Bottom right) swab placed in jar and sterilized.

negative in the absence of CPE after two successive passages in tissue cultures. When viruses were present they were identified by neutralization with type-specific antisera (Wenner et al., 1954).

C. Results

The poliomyelitis epidemic in Des Moines (Chin and Gravelle, 1961; Chin et al., 1961) began in the latter part of May and ended in the first part of October. The cases appeared in two distinct peaks, one in late June, one in late August. The first part of the epidemic consisted mainly of paralytic poliomyelitis, with the cases concentrated in the central, crowded areas of the city; the second half consisted predominantly of nonparalytic cases scattered in different areas of the city.

From the first week of July to the end of September, weekly sewage samples were obtained from the eight sampling areas. A total of 94 samples was taken. The percentage of samples containing poliovirus 1 varied from 9.1% (2 in 22) in the upper socioeconomic areas to 100% in the lower middle socioeconomic areas (Table 2). In the lower socioeconomic areas, all samples obtained during the first part of the epidemic contained poliovirus; but in the second half of the epidemic, when the number of cases in lower socioeconomic areas diminished, polioviruses were isolated from only 50% of the samples. The relationship, by socioeconomic classification, between the attack rates of poliomyelitis and the prevalence of poliovirus in sewage is shown in Table 3. It is evident that the prevalence of poliovirus in sewage is a reflection of the incidence of poliomyelitis in the community.

From July 6 to August 12, 34 different viruses were recovered from the sewage samples; all but one were poliovirus. In the later part of the epidemic, 6 of 32 viruses recovered were not poliovirus. During the comparable period in the first part of the epidemic (July 2 to August 12), 37 reported cases of viral infection were etiologically confirmed: 36 were caused by poliovirus and one by a "non-poliovirus." In the second part of the epidemic, 15 of 26 confirmed cases were caused by poliovirus, while 11 were caused by non-polioviruses (Table 3). Although the number of non-polioviruses recovered from the sewage in the latter half of the epidemic was small, the finding in sewage did give an indication of the relative incidence of poliovirus and non-poliovirus infection occurring in the community.

Attempts were made to estimate the concentration of virus in sewage samples. Infectivity titrations in monkey kidney tissue cultures were performed on nine samples that contained virus. The titres varied from less than $10^{0.3}$ to $10^{0.8}$ 50% tissue culture infective doses ($TCID_{50}$) per ml of sewage; the

TABLE 2

ISOLATION OF POLIOVIRUS 1 FROM WEEKLY SEWAGE SAMPLES IN EIGHT
SELECTED AREAS DURING THE POLIMYELITIS EPIDEMIC IN DES MOINES,
IOWA, JULY 6 TO SEPTEMBER 28, 1959

| Sampling areas | Socio-economic classification | Frequency of poliovirus 1 isolation from weekly samples | | |
		7/6 to 8/10	8/17 to 9/28	Total
5,7	Upper	1/12 [a]	1/10	2/22
1,7	Upper middle	7/12	5/12	12/24
2,4	Lower middle	12/12	12/12	24/24
3,8	Lower	12/12	6/12	18/24
From 8 sampling areas		32/48	24/46	56/94

[a] $\frac{\text{Numerator}}{\text{Denominator}} = \frac{\text{Number of samples positive}}{\text{Number of samples tested}}$.

TABLE 3

RELATION BY SOCIO-ECONOMIC CLASSIFICATION, BETWEEN INCIDENCE
OF POLIOMYELITIS AND FREQUENCY OF POLIOVIRUS RECOVERY FROM
SEWAGE IN DES MOINES, IOWA, 1959

| Socio economic classification | Estimated population | Reported poliomyelitis | | Frequency of poliovirus 1 recovery from sewage | |
		No. of cases	Rate/ 100,000	Ratio [a]	%
Upper	64,100	10	15.6	2/22	9.1
Middle	97,100	50	51.5	36/48	75.0
Lower	41,900	75	179.0	18/24	75.0
Total	203,100	135	66.5	56/94	59.6

[a] $\frac{\text{Numerator}}{\text{Denominator}} = \frac{\text{Number of samples positive}}{\text{Number of samples tested}}$.

mean was $10^{0.4}$ TCID$_{50}$. The average infectivity titre of poliovirus 1 recovered from the sewage by the gauze pad technique therefore was about 250
TCID$_{50}$/100 ml of original sewage.

Consideration of the above study shows that the gauze pad technique is a
highly efficient method for sampling viruses in sewage. During oral polio-
vaccine campaigns, poliovirus (vaccine virus) can be readily detected in sewage, and the sensitivity of the method may be illustrated. For example, in
a study done by Gelfand et al. (1962), poliovirus was recovered from sewage
2 to 3 months after the vaccine was given.

III. BACTERIOPHAGE PLAQUE METHOD

A. General Considerations

A bacteriophage is a virus that reproduces entirely in the bodies of living
bacteria. In its parasitic action, the bacterophage uses the host bacterium to
produce additional viral particles. The host cell is killed in the process. The
process of this multiplication at the expense of the bacterial bodies is called
bacteriophagy. The effect of bacteriophagy leads finally to a lysis of the bacterial cells, a phenomenon that in dense cultures manifests itself to the naked
eye as a clearing of the bacterial culture. In agar plating, described on p. 382,
"plaques" or lysed areas on the agar may be seen interspersed between areas
of solid bacterial growth (Fig. 3). These plaques represent "colonies" of bacteriophage particles. In the process of bacteriophagy, a bacterial cell adsorbs
the phage particle. Multiplication of the phage takes place in the host until the
cell ruptures, liberating a number of phage particles. The newly formed phage
in turn infect other bacteria. A chain reaction of infection and lysis of bacteria
is thus initiated, resulting in the formation of a plaque. The reaction ceases
when the bacterial growth reaches the stationary phase. The plaque then reaches maturity and does not further increase in size.

B. Method of Isolation from Water Sources

Fifty milliliters of broth is placed in a 200-ml flask. One milliliter of
visibly turbid 2-4 hr broth culture of the host bacterium against which the
bacteriophage is to be isolated is added. If sewage water is being tested it
should be centrifuged for a few minutes to remove other bacteria and fungi.
Two milliliters of the supernatant is added to the flask. The mixture is then
incubated at 37°C for 12-18 hr. The broth is then centrifuged and passed

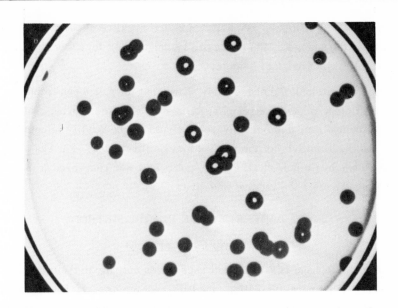

FIG. 3. Plaques formed on sensitive cells by plating samples of cultures of lysogenic <u>Bacillus</u> <u>megatherium</u>. Clear plaques are formed by colonies that liberate phage during growth.

through a bacterial filter to remove bacteria. The bacteriophage, if present, passes through the filter and is present in the filtrate.

In testing river or pond water where phage may be present in lower concentrations, an enrichment method is preferred. Ten milliliters of 20% peptone water is added to 190 ml of water in a 500-ml flask. Four percent sterile NaOH solution is used to adjust the pH to a value that is most suitable for the growth of the bacterium for which the bacteriophage is to be isolated. From this point the procedure is the same as for sewage water.

C. Assay by Local Lesion Count

The classic example of a local lesion type of virus assay is the highly accurate plaque count of bacterial viruses. The procedure as it is used currently is presented in detail by Adams (1950) and is carried out as follows. Dilutions of virus are mixed with a bacterial suspension in nutrient agar and poured over the surface of ordinary agar plates. Clear areas or plaques are produced by the virus in the confluent growth of bacteria in the agar overlay after a suitable incubation period at a temperature suited for phage and bacteria. The plaque count per plate divided by the volume and dilution of the virus inoculum gives the titre in plaque-forming units per milliliter of original suspension.

A direct proportionality is observed between plaque count and relative virus concentration when both variates are plotted on the same scale (usually logarithmic); thus the curve is linear with a slope of 1. Because of this linear relationship, each plaque is considered an infective center initiated by a single virus particle infecting a susceptible bacterium. The plaque count does not necessarily represent the total virus particle count but only that fraction capable of absorbing to and infecting viable bacterial cells. This fraction is called the "efficiency of plating" and can be estimated in a relative sense by comparing the plaque count with the infectivity of the virus for bacterial cells in broth suspension where conditions are usually more favorable for virus absorption. For example, a bacterial virus preparation in high dilution is added to a suspension of susceptible bacteria. Small aliquots of the mixture, designed to yield on the average less than one infected bacterium per sample, are incubated until the infected bacteria have burst, liberating viruses.

Each aliquot is plated and the number of plates yielding no plaques as well as those yielding large numbers of plaques is noted. From the proportion of plates with no plaques one can estimate the average number of virus particles n_1 per aliquot from the Poisson formula

$$P_o = \underline{e}^{-n_1}$$

where P_o is the fraction of samples containing no infective virus particle and $\underline{e}$ is the base of natural logarithms. A parallel titration of the stock bacterial virus preparation by the usual plaque assay technique will also yield an average value n_2 for the number of infective particles per sample. The ratio n_2/n_1, the "efficiency of plating" coefficient, is usually less than 1 and indicates that fraction of the infected bacteria in suspension that will go on to produce plaques after plating (Ellis and Delbruck, 1939).

Absolute efficiency of plating can only be estimated after some independent measurement of the concentration of characteristic physical particles has been made (Luria, 1953).

The precision of assay by the bacteriophage plaque technique is readily estimated since the variance of a Poisson distribution equals the mean (Isaacs, 1957). For an average plate count of n plaques, the standard deviation equals

$\sqrt{n}$ which, divided by the average count n and multiplied by 100, yields a value for the coefficient of variation. For example, an average plaque count of 100 per plate has a coefficient of variation of 10%, while 25 plaques per plate yield a standard deviation that is 20% of the count. The higher the count per plate, the greater the precision of assay within the limits imposed by the size of the agar plate.

D. Agar Plating of Bacteriophage

The agar method for plating bacterial viruses was first described by Gratia (1936) and is in general use today by practically all workers.

The host bacteria and virus particles are mixed in a small volume of warm 0.7% agar and the mixture is poured over the surface of an ordinary agar plate and allowed to harden to form a thin layer. This thin layer is actually poured on a deep layer of prepoured 1.5% agar, which is used as foundation. The plaques appear as circular areas in the opaque layer of bacterial growth.

For the plating procedure the soft 0.7% agar is melted in a boiling water bath and cooled in a 46°C water bath. It is then transferred with a warmed pipet in 2.5 ml amounts to warmed test tubes in the 46°C water bath. The bacterial inoculum is prepared by washing the bacteria from the surface of an agar slant with 5 ml of broth. One drop of the resultant suspension is added to each tube of soft agar. The diluted virus in any volume up to 1 ml is then pipeted into the tubes of soft agar and the mixture poured immediately over the surface of an agar plate. The plate is rocked gently to give an even mixture of the bacteria and viral particles on the surface. Both the 1.5% agar and the soft agar layer should be allowed to harden with the Petri dish resting on a leveled sheet of plate glass to ensure uniform distribution of the bacteriophage and bacteria over the plate surface. If the melted soft agar is held at 46°C for much over an hour the agar will start to gel and poor plates will result. Neither bacteria nor viruses seem to be harmed.

E. Plaque Counting and Calculation of Titre

The localized lesions, nearly always enhanced in contrast with a suitable stain, are usually counted with the unaided eye. Some lesions may have to be counted microscopically. To count microscopically, areas of dead cells (presumably plaques) can be distinguished from gaps in the cell sheet by their

gray appearance when held against a black background; most virus strains
produce round and symmetrical plaques, and in this case areas of dead cells
with irregular outlines indicate a nonspecific cause of cell death. An automatic recording pen that does not depend on electrodes immersed in the
medium is available commercially (Scientifica, London, England).

Plaque counts are recorded only for those plates in which losses by overlapping of plaques are statistically negligible (Cooper, 1961). An adequate
guide is that 100 plaques per plate will give no significant loss by overlapping
if the internal diameter of the Petri plate is at least 25 times the average
plaque diameter.

The virus titre is obtained by multiplying the average plaque count per
plate by the total dilution factor. Thus an average of 66 plaques per plate
obtained from plating 0.2 ml of a 3×10^{-4} dilution represents a titre in the
original preparation of $66/0.2 \times 3 \times 10^{-4} = 1.0 \times 10^{7}$ plaque-forming units
per milliliter.

This method of plaque counting and the calculation of titre may also be
applied to animal viruses (see next section).

IV. MONOLAYER KIDNEY TISSUE CULTURE

A. Principles of Tissue Culture

The obligatory parasitic nature of animal viruses means that living cells
must be made available for their isolation, growth, and general study. The
principles underlying tissue culture procedures are therefore concerned with
the provision and maintenance of suitable cells in such a way that they are able
to support virus replication. The procedures usually involve growing cells
as monolayer sheets upon glass surfaces and maintaining them in appropriate
nutrient media so that they remain viable for a prolonged period. Other techniques provide for the similar maintenance of single cells in isolation and of
large numbers in bulk fluid suspension. By the use of yet other procedures,
monolayers can be maintained under solid nutrient agar media.

The frequent observation that animal viruses propagating in monolayer
cultures of mammalian cells produced a grossly visible cytopathic effect was
quickly exploited by Dulbecco (1952) for the development of a plaque-count
assay. This opened the door to quantitative studies with animal viruses
equivalent in precision to those made with bacterial viruses. Thus bacterial
and animal virus plaque-count assays became analogous in principle.

B. Trypsinized Cell Culture of Dulbecco

An important technique was developed by Dulbecco and Vogt (1954) for the precise determination of the number of infective particles of a cytopathogenic virus that may be present in a given suspension. The method consists in establishing a continuous monolayer of monkey tissue cells at the bottom of a Petri dish. The cells are then exposed to a high dilution of the suspension of virus. After a short interval they are covered with a nutrient agar mixture that serves to localize the virus in the areas in which infection of individual cells has taken place. Multiplication of virus is followed by destruction of adjacent cells. This effect becomes visible within 24-48 hr and resembles the plaques formed by bacteriophage in confluent agar cultures of bacteria.

C. Trypsinization Process

This method is based on the depolymerization of the intercellular ground substance of a tissue by a trypsin solution at pH 7.4-8. When fragments of tissue are immersed in a trypsin solution, cells detach from each other and a suspension of single highly viable cells is obtained.

The procedure, as modified by Younger (1954), is summarized here. The cortical area of a kidney from a Rhesus or Cynomolgus monkey is minced (Fig. 4). The minced tissue, after being washed in buffer solution, is treated with trypsin solution at $37^{\circ}C$. After a short interval the solution is discarded and fresh solution added. Digestion is allowed to proceed for a short length of time while the suspension of tissue is gently agitated at a low speed in a Waring blender. Mixing is controlled by a rheostat. After the fragments have settled, the supernatant fluid is removed and the coarse material it contains is separated by filtration through gauze. This procedure is repeated until nearly all the cells have been separated from the tissue. The aliquots of trypsinized suspension are pooled and centrifuged and the cells washed with the nutrient medium employed for their cultivation.

From the packed cells a suspension is prepared, to be used for standardization. This is accomplished by determining the number of cells by counts of the nuclei or of the optical density. For use in tube cultures, the suspension is adjusted to contain 600,000 cells per ml. One-half milliliter of this suspension is added to a tube, which is stoppered and incubated in a nearly horizontal position. The cells settle on the glass and increase in size, ultimately forming a continuous layer. After six or seven days the cultures are ready

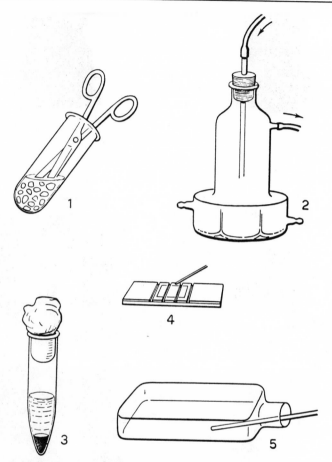

FIG. 4. Phases of trypsinization process: (1) Fragmentation of tissue;
(2) Trypsinization of the fragments; (3) Sedimentation of the cells; (4) Counting of the viable cells; (5) Distribution of cells suspended in growth medium.

for inoculation. From 800 to 1000 cultures have been prepared in this way
from the renal tissue of one monkey. Using these cell suspensions, cultures
in flat-sided bottles (40 mm x 110 mm) have been successfully established
and applied to the propagation of larger quantities of virus (Fig. 5).

D. Titration

As has been found with bacterial virus plaque assays, a linear relationship exists between plaque count and relative virus concentration for most of
the animal viruses assayed by this method thus far. On the basis of this proportionality, Dulbecco and Vogt (1954) have established on statistical grounds

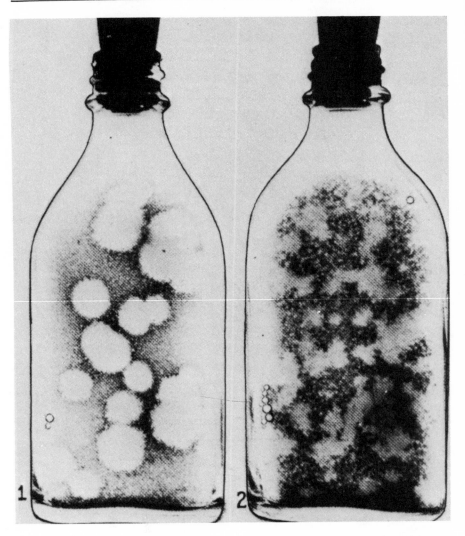

FIG. 5. Plaques of type III poliovirus (1) and type 6 ECHO virus (2), in a thin layer of monkey epithelial cells growing on a special nutrient medium solidified with agar.

that a single virus particle is sufficient to produce a plaque (see Bacterio-phage Plaque Method, p. 382).

E. Assay Based on All-or-None Response

The end-point dilution method of assay based upon an all-or-none response is used most commonly for the titration of animal viruses. Usually, a standard volume of serial tenfold or fivefold dilutions of a virus preparation is inoculated into test groups of five or six susceptible animal or tissue culture hosts and the number of positive and negative responses scored for each dilution. A positive response may be death as a result of infection or some gross manifestation of infection, such as paralysis. In the case of infected tissue culture, where each culture is analogous to an individual animal host, readily observable cytopathic effects, such as complete cellular degeneration or giant cell formation, may represent a positive score. A plot of the percentage incidence of positive scores as a function of the log dilution of virus suspension will yield a sigma-shaped dose-response curve. By interpolation, one can estimate the dilution at which 50% positive and 50% negative responses occur. This is called the 50% infectivity end-point dilution (ID_{50}) and represent that point on the curve where the slope is steepest, i.e., where the smallest change in virus concentration will produce the greatest difference in response. The ID_{50} is therefore the most precisely measured point on the curve. Relative virus concentrations are calculated from the inoculum volume and dilution and are expressed as the number of ID_{50} per milliliter of original virus suspension.

388 WATER POLLUTION: Disposal and Reuse

REFERENCES

Adams, M. H. 1950. "Methods of study of bacterial viruses." Methods in Med. Res. 2, 1-74.

Baron, S. 1957. "Ultracentrifuge concentration of polio virus and effect of calf serum and gelatin. Proc. Soc. Exptl. Biol. Med. 96, 760-764.

Benson, L. M. and J. E. Hotchin. 1960. "Cytopathogenicity and plaque formation with lymphocytic choriomeningitis virus." Proc. Soc. Exptl. Biol. Med. 103, 623-625.

Chin, T. D. Y. and C. R. Gravelle. 1961. "Enterovirus isolations from sewage." J. Inf. Diseases 109, 205-209.

Chin, T. D. Y., W. M. Marine, E. C. Hall, C. R. Gravelle, and J. F. Speers. 1961. "Poliomyelitis in Des Moines, Iowa." Am. J. Hyg. 74, 67-94.

Cooper, P. D. 1961. "The plaque assay of animal viruses." Adv. Virus. Res. 8, 319-378.

Dulbecco, R. 1952. "Production of plaque in monolayer tissue cultures by single particles of animal virus." Proc. Natl. Acad. Sci. U.S. 38, 747-753.

Dulbecco, R. and M. Vogt. 1954. "Plaque formation and isolation of pure lines with poliomyelitis viruses." J. Exptl. Med. 99, 167-182.

Ellis, E. L. and M. Delbrück. 1939. "The growth of bacteriophage." J. Gen. Physiol. 22, 364-384.

Enders, J. F. 1952. "General preface to studies on the cultivation of poliomyelitis viruses in tissue culture." J. Immunol. 69, 639-643.

Gelfand, H. M., A. H. Holguin and R. A. Feldman. 1962. "Community-wide type 3 oral poliovirus vaccination in Atlanta, Ga." J. Am. Med. Assoc. 181, 281-289.

Gratia, A. 1936. "Numerical relations between lysogenic bacteria and particles of bacteriophage." Ann. Inst. Pasteur 57, 652-676.

Isaacs, A. 1957. "Particle counts and infectivity titrations for animal viruses." Adv. Virus Res. 4, 111-158.

Luria, S. E. 1953. General Virology, Wiley, New York.

MacCallum, F. O., W. C. Cockburn, E. H. R. Smithard and S. L. Wright. 1952. "The use of gauze swabs for the detection of poliomyelitis virus in sewers. 2nd International Poliomyelitis Conference, 484-487.

Moore, A. 1948. "The detection of paratyphoid carriers in towns by means of sewage examination." Month. Bull. Min. Health 7, 241-248.

Wenner, H. A., C. A. Miller, P. Kamitsuka, and J. C. Wilson. 1954. "Preparation and standardization of antiserum." Am. J. Hyg. 59, 221-235.

Younger, J. S. 1954. "Monolayer tissue cultures. I. Preparation and standardization of suspensions of trypsin-dispersed monkey kidney cells." Proc. Soc. Exptl. Biol. Med. 85, 202-205.